Principles of
Foundation Engineering

Principles of Foundation Engineering

Rey Hendricks

Larsen & Keller
www.larsen-keller.com

Principles of Foundation Engineering
Rey Hendricks
ISBN: 978-1-64172-118-9 (Hardback)

 **Larsen & Keller**

Published by Larsen and Keller Education,
5 Penn Plaza,
19th Floor,
New York, NY 10001, USA

Cataloging-in-Publication Data

Principles of foundation engineering / Rey Hendricks.
    p. cm.
Includes bibliographical references and index.
ISBN 978-1-64172-118-9
1. Foundations. 2. Structural engineering. I. Hendricks, Rey.
TA775 .P75 2019
624.15--dc23

For more information regarding Larsen and Keller Education and its products, please visit the publisher's website www.larsen-keller.com

Table of Contents

Permissions

Index

Preface

Foundation engineering is a branch of engineering that applies the principles of soil and rock mechanics for the design of the foundational elements of architectural structures. A foundation connects a structure with the ground and is responsible for the transference of the structural load to the ground for stability of the structure. Some historic foundation designs are stone foundations, rubble trench foundations, padstones and post in ground construction. Modern foundations can be either shallow or deep. Shallow foundations include spread footing and slab-on-grade foundation. Deep foundations include drilled shafts, Earth stabilized columns, impact driven piles, etc. This book elucidates the concepts and innovative models around prospective developments with respect to foundation engineering. While understanding the long-term perspectives of the topics, the book makes an effort in highlighting their impact as a modern tool for the growth of the discipline. It aims to serve as a resource guide for students and experts alike and contribute to the growth of the discipline.

To facilitate a deeper understanding of the contents of this book a short introduction of every chapter is written below:

Chapter 1, Foundation engineering is the application of rock mechanics and soil mechanics in the design of the structural elements of a foundation. This chapter has been carefully written to provide an introduction to the varied aspects of foundation and foundation engineering, and includes topics like civil engineering, geotechnical engineering and geomechanics. **Chapter 2**, A foundation is an architectural element which connects a structure to the ground. This leads to the easy transference of load. There are various kinds of shallow and deep foundations. This chapter closely examines such types of foundations, such as rubble trench foundation, shoring, underpinning and monopole foundation, among others. **Chapter 3**, The study of the behavior of soils, or analysis of the flow of fluids in natural and man-made structures as well as investigation of the deformation of structures that are supported by or made of soil are within the scope of soil mechanics. The aim of this chapter is to provide an insight into the field of soil mechanics, through topics such as frost heaving, soil shrinking swelling, bearing capacity, stresses in the ground, lateral earth pressures, soil porosity, soil degradation, etc. **Chapter 4**, Rock mechanics is an applied and theoretical science concerned with the study of the mechanical behavior of rocks, such as the influence of force fields on rock masses. An important technique for rock mass classification and rock slope engineering is the Q-slope method. All the diverse principles of rock mechanics and Q-slope have been carefully examined in this chapter. **Chapter 5**, Machine foundations are special foundations that are required for machine tools, machines and heavy equipment, which operate under varied loads, speeds and conditions. Such foundations are designed as per the dynamic forces that result from the operation of these machines. The topics elaborated in this chapter on vibration, machine foundation analysis and design, vibration analysis of machine foundation, design charts for machine foundation, damping, etc. will help in providing a comprehensive understanding of machine foundations.

Finally, I would like to thank the entire team involved in the inception of this book for their valuable time and contribution. This book would not have been possible without their efforts. I would also like to thank my friends and family for their constant support.

Rey Hendricks

Introduction to Foundation Engineering

Foundation engineering is the application of rock mechanics and soil mechanics in the design of the structural elements of a foundation. This chapter has been carefully written to provide an introduction to the varied aspects of foundation and foundation engineering, and includes topics like civil engineering, geotechnical engineering and geomechanics.

Foundation

Foundation is the lowest part of the building or the civil structure that is in direct contact with the soil which transfers loads from the structure to the soil safely. Generally, the foundation can be classified into two, namely shallow foundation and deep foundation.

A shallow foundation transfers the load to a stratum present in a shallow depth. The deep foundation transfers the load to a deeper depth below the ground surface.

A tall building like a skyscraper or a building constructed on very weak soil requires deep foundation. If the constructed building has the plan to extend vertically in future, then a deep foundation must be suggested.

Figure: Building Foundation in Construction

To construct a foundation, trenches are dig deeper into the soil till a hard stratum is reached. To get stronger base foundation concrete is poured into this trench. These trenches are incorporated with reinforcement cage to increase the strength of the foundation.

The projected steel rods that are projected outwards act as the bones and must be connected with the substructure above. Once the foundation has been packed correctly the construction of the building can be started.

The construction of the foundation can be done with concrete, steel, stones, bricks etc. The material and the type of foundation selected for the desired structure depends on the design loads and the type of underlying soil.

The design of the foundation must incorporate different effects of construction on the environment. For example, the digging and piling works done for deep foundation may result in adverse disturbance to the nearby soil and structural foundation. These can sometimes cause the settlement issues of the nearby structure.

Such effects have to be studied and taken care before undergoing such operations. Disposal of the waste material from the operations must be disposed properly. The construction of foundation has to be done to resist the external attack of harmful substances.

The foundation for each structure is designed such that:

- The underlying soil below the foundation structure does not undergo shear failure.

- The settlement caused during the first service load or have to be within the limit.

- Allowable bearing pressure can be defined as the pressure the soil can withstand without failure.

Purpose of Foundation

Foundations are provided for all load carrying structure for following purposes:

- Foundation is the main reason behind the stability of any structure. The stronger is the foundation, more stable is the structure.

- The proper design and construction of foundations provide a proper surface for the development of the substructure in a proper level and over a firm bed.

- Specially designed foundation helps in avoiding the lateral movements of the supporting material.

- A proper foundation distributes load on to the surface of the bed uniformly. This uniform transfer helps in avoiding unequal settlement of the building. Differential settlement is an undesirable building effect.

- The foundation serves the purpose of completely distributing the load from the structure over a large base area and then to the soil underneath. This load transferred to the soil should be within the allowable bearing capacity of the soil.

Functions of Foundation in Construction

Based on the purposes of foundation in construction, the main functions of the foundation can be enlisted as below:

1. Provide overall lateral stability for the structure

2. Foundation serve the function of providing a level surface for the construction of substructure

3. Load Distribution is carried out evenly

4. The load intensity is reduced to be within the safe bearing capacity of the soil

5. The soil movement effect is resisted and prevented

6. Scouring and the undermining issues are solved by the construction of foundation.

Requirements of a Good Foundation

The design and the construction of a well-performing foundation must possess some basic requirements that must not be ignored. They are:

1. The design and the construction of the foundation is done such that it can sustain as well as transmit the dead and the imposed loads to the soil. This transfer has to be carried out without resulting in any form of settlement that can result in any form of stability issues for the structure.

2. Differential settlements can be avoided by having a rigid base for the foundation. These issues are more pronounced in areas where the superimposed loads are not uniform in nature.

3. Based on the soil and area it is recommended to have a deeper foundation so that it can guard any form of damage or distress. These are mainly caused due to the problem of shrinkage and swelling because of temperature changes.

4. The location of the foundation chosen must be an area that is not affected or influenced by future works or factors.

Foundation Engineering

Foundation engineering, applies theoretical knowledge concerning the behavior of soils and rocks and the construction of load-bearing structures to the planning and construction of foundations for infrastructure. At the most basic level, a foundation engineer would consider the kind of soil on which construction is to begin, allowing for the selection of the best material for the job, taking into account variables such as the manner in which such materials would need to be reinforced. This field of engineering not only establishes the physical qualities and quantities needed for the construction of foundations but establishes the necessary design parameters needed for such construction. Such parameters are established by evaluating factors such as the bearing capacity of a particular soil, allowable soil pressure, and the influence of slopes and adjacent foundations, among others.

Foundation Engineering is Critical in Construction

An equally important facet of foundation engineering entails the maintenance and evaluation of existing foundations, which in practice involves pre-empting the degradation or failure of a foundation, and in some cases assessing the damage that has already occurred. Such risk assessments require the foundation engineer to take into account factors such as the arrangement of physical features, or topography, of the area being assessed, seismic forces, and groundwater, all of which may play a part in the deterioration of existing foundations. This aspect of foundation engineering is particularly important in minimizing risks to human safety which may occur as a result of unsound foundations. Extrapolating along the lines of risk assessment, site conditions which may have otherwise limited development potential may be mitigated through the improvement of the engineering properties of the soil and rock foundations in themselves.

Simply described, the foundation of a structure serves to transmit loads from the structure to the earth. As such, the design of a foundation requires an estimation of the magnitude and location of loads that need supporting, a plan for the evaluation of the subsurface, and the establishment of the required soil parameters through field testing. Once these factors have been taken into consideration, the foundation is designed such that construction is achieved in the most economic manner feasible, while also ensuring that any risks which may be present during, or subsequent to the construction of the foundation, are minimized.

Importance of Soil in Foundation Engineering

The most significant factors to be taken into consideration when designing foundations include those concerning settlement and bearing capacity. Settlement takes into account the tendency for soils to undergo consolidation, a process whereby soils decrease in volume over time under permanent loads – in the case of foundation engineering, the weight of the structure and the foundation itself. This process occurs due to the expulsion of water from soil without any concomitant replacement with water or air, resulting therefore in an overall decrease in the volume of soil. This was probably, at least, one of the variables the designers of the Tower of Pisa neglected to consider, which eventually resulted in the distinctive tilt which gives the Tower its more recognizable label as the Leaning Tower of Pisa.

Bearing capacity refers to the property of soil to support forces applied to the ground. Strictly defined, the bearing capacity of soil would refer to the maximum contact pressure, averaged over time, between the foundation and the soil, which would not produce shear failure in the soil which itself is related to the shear strength of the soil. This variable, in turn, is determined by the friction and interlocking between soil particles.

The history of foundation engineering may be traced as far back as Ancient Greece. As cities grew larger in parallel with the progress of those ancient civilizations, buildings were erected on foundations which were specifically constructed to serve as supporting structures. Until the 18th century, however, no formalized discipline concerning the theoretical underpinnings of soil design existed, and construction proceeded based on past experience rather than any demonstrable protocol. The lack of anything resembling a scientific approach to the matter of foundation engineering soon manifested itself in such famous engineering problems as the Leaning Tower of Pisa, itself the result of an inadequate foundation on soil too soft on one side to completely support the weight of the tower. These problems soon prompted scientists to begin examining the subsurface in a more systematic fashion, resulting in the progression of foundation engineering from century through to 1925, when modern foundation engineering is said to have begun, the 18th with the publication of Erdbaumechanick by the father of soil mechanics, Karl von Terzaghi.

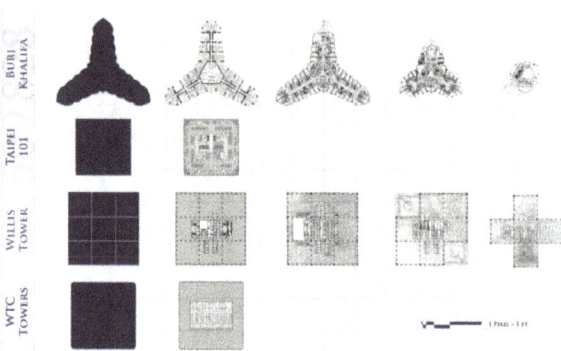

Historically, foundations simply consisted of buildings or structures being built in contact with the ground, or otherwise in a manner known as 'post in ground' construction. Post in ground, also known as earthfast construction, simply consisted of vertical roof-bearing posts being placed into excavated post holes, and has been in use from the Neolithic period to the present. Structures utilizing such construction methods are relatively impermanent but are simply to erect. Padstones are another type of foundation which has been in use since Ancient Greece, and is simply single stones which both spread the weight of a structure on the ground as well as elevating the structure. Such elevation was of particular use in preventing the contents of granaries from vermin as well as water seepage.

Depth of Foundations

Spread footing foundations, a type of shallow foundation, are typically used in residential buildings, and consists of a wider lower portion which supports the load-bearing foundation walls, such as to distribute the weight of the building over a greater area for increased stability. The design and implementation of spread footings is influenced by soil factors, as well as the weight of the structure that is being supported – this said, it could be argued that the design of nearly any foundation which is placed, shallow or deep, would have to take into account these two factors. Spread footings may be adapted to use in sites where gradients exist by being constructed as a series of steps cut into the gradient. Such use of a spread footing is called a stepped footing. Another example of shallow foundation previously mentioned is the slab-on-grade foundation, also known as floating slab foundations. These foundations are constructed by allowing a concrete slab to set within a mold that is set into the ground; this concrete slab then serves as the foundation. The advantages of this type of foundation are that it is cost-effective and relatively sturdy, however these points have to be balanced against the fact that such a foundation restricts underground access to utility lines and can result in significant heat loss when ground temperatures fall lower than that of indoor temperatures. This second drawback may be mitigated however, by the use of insulation or heating systems, although these solutions may entail additional costs/issues in themselves.

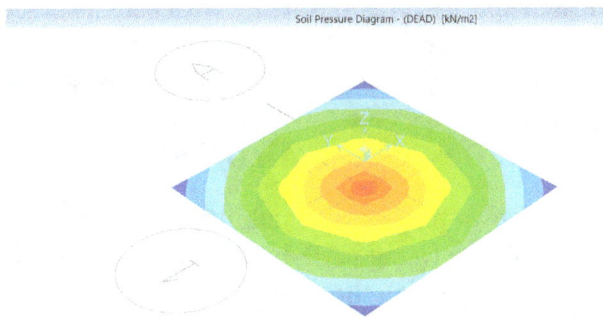

Deep foundations may be recommended over shallow foundations for a number of reasons, but these reasons most commonly tend to be due to exceedingly large structural loads, such as those that would be present in the construction of high rise buildings, poor quality soil at superficial depths, or a lack of space for a sufficiently large shallow foundation to be able to support a structure properly. As mentioned above, deep foundations may be classified according to the type of foundation itself, the method by which the foundation is placed (also related to the type of foundation) as well as the material the foundation is constructed from. Common materials used as deep foundations include reinforced concrete, prestressed concrete, timber and steel. Deep foundations which are driven into the ground are typically done so using a pile driver, a machine which operates by alternatingly raising and releasing a weight onto a pile until it has reached the required depth.

Importance of Piles

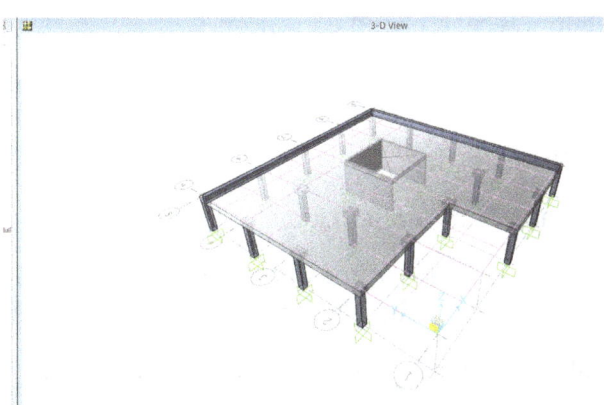

A particular advantages associated with driven piles are that the soil which is displaced during the pile-driving process becomes compressed, effectively, acting as a splint, and thereby increasing the load-bearing capacity of the pile. Furthermore, driven piles are considered to have been 'evaluated' for their weight bearing ability, as the weight required to drive the pile into the ground must necessarily exceed that of the structure which the pile is supporting, therefore offering a considerable degree of certainty that the pile will be able to serve its function. Drilled piles make use of the largest diameter piles, and permit pile placement even in dense subsurface soil or rock. Examples of drilled piles include under reamed piles, used in softer strata, as well as Augercast piles, which are often used when it is necessary to minimize noise pollution as well as in environmentally sensitive sites. These unique piles are formed using hollow stemmed augers which drill down to a predetermined depth, following which cement if flowed down through the stem, and the auger is withdrawn, allowing the drilled space to fill up directly with cement. As the auger is withdrawn, a column of fluid cement is formed within the drilled site, into which reinforcing cages are commonly placed to ensure that the pile is strong enough to serve its purpose.

Importance of Foundation Design and Analysis

Each structural challenge has its own parameters making foundation design and analysis a crucial element for all construction projects. Using CAD-based tools, Civil Engineers test these parameters to find out the correct way to approach the project.

Civil Engineering

Civil Engineering is one of the broadest and oldest of the engineering disciplines, extending across many technical specialties. Civil engineering is everything you see that's been built around us. It's about roads and railways, schools, offices, hospitals, water and power supply and much more. The kinds of things we take for granted but would find life very hard to live without.

Civil Engineer

Civil engineers conceive, design, build, supervise, operate, construct and maintain infrastructure projects and systems in the public and private sector, including roads, buildings, airports, tunnels, dams, bridges, and systems for water supply and sewage treatment. Many civil engineers work in planning, design, construction, research, and education.

Duties of Civil Engineers

Civil engineers typically do the following:

- Analyze long range plans, survey reports, maps, and other data to plan and design projects.

- Consider construction costs, government regulations, potential environmental hazards, and other factors during the planning and risk-analysis stages of a project.

- Compile and submit permit applications to local, state, and federal agencies, verifying that projects comply with various regulations.

- Oversee and analyze the results of soil testing to determine the adequacy and strength of foundations.

- Analyze the results of tests on building materials, such as concrete, wood, asphalt, or steel, for use in particular projects.

- Prepare cost estimates for materials, equipment, or labor to determine a project's economic feasibility.

- Use design software to plan and design transportation systems, hydraulic systems, and structures in line with industry and government standards.

- Perform or oversee surveying operations to establish building locations, site layouts, reference points, grades, and elevations to guide construction.

- Manage the repair, maintenance, and replacement of public and private infrastructure.

Civil engineers also must present their findings to the public on topics such as bid proposals, environmental impact statements, or property descriptions.

Many civil engineers hold supervisory or administrative positions ranging from supervisor of a construction site to city engineer, public works director, and city manager. As supervisors, they are tasked with ensuring that safe work practices are followed at construction sites.

Other civil engineers work in design, construction, research, and teaching. Civil engineers work with others on projects and may be assisted by civil engineering technicians.

Civil engineers prepare permit documents for work on projects in renewable energy. They verify that the projects will comply with federal, state, and local requirements. These engineers conduct structural analyses for large-scale photovoltaic, or solar energy, projects. They also evaluate the ability of solar array support structures and buildings to tolerate stresses from wind, seismic activity, and other sources. For large-scale wind projects, civil engineers often prepare roadbeds to handle large trucks that haul in the turbines.

Civil engineers work on complex projects, and they can achieve job satisfaction in seeing the project reach completion. They usually specialize in one of several areas.

Construction engineers manage construction projects, ensuring that they are scheduled and built in accordance with plans and specifications. These engineers typically are responsible for the design and safety of temporary structures used during construction. They may also oversee budgetary, time-management, and communications aspects of a project.

Geotechnical engineers work to make sure that foundations for built objects ranging from streets and buildings to runways and dams are solid. They focus on how structures built by civil engineers, such as buildings and tunnels, interact with the earth (including soil and rock). In addition, they design and plan for slopes, retaining walls, and tunnels.

Structural engineers design and assess major projects, such as buildings, bridges, or dams, to ensure their strength and durability.

Transportation engineers plan, design, operate, and maintain everyday systems, such as streets and highways, but they also plan larger projects, such as airports, ship ports, mass transit systems, and harbors.

Important Qualities of Civil Engineers

Decision making skills: Civil engineers often balance multiple and frequently conflicting objectives, such as determining the feasibility of plans with regard to financial costs and safety concerns. Urban and regional planners often look to civil engineers for advice on these issues. Civil engineers must be able to make good decisions based on best practices, their own technical knowledge, and their own experience.

Leadership skills: Civil engineers take ultimate responsibility for the projects that they manage or research that they perform. Therefore, they must be able to lead planners, surveyors, construction managers, civil engineering technicians, civil engineering technologists, and others in implementing their project plan.

Math skills: Civil engineers use the principles of calculus, trigonometry, and other advanced topics in mathematics for analysis, design, and troubleshooting in their work.

Organizational skills: Only licensed civil engineers can sign the design documents for infrastructure projects. This requirement makes it imperative that civil engineers be able to monitor and evaluate the work at the jobsite as a project progresses. That way, they can ensure compliance with the design documents. Civil engineers also often manage several projects at the same time, and thus must be able to balance time needs and to effectively allocate resources.

Problem-solving skills: Civil engineers work at the highest level of the planning, design, construction, and operation of multifaceted projects or research. The many variables involved require that they possess the ability to identify and evaluate complex problems. They must be able to then use their skill and training to develop cost-effective, safe, and efficient solutions.

Speaking skills: Civil engineers must present reports and plans to audiences of people with a wide range of backgrounds and technical knowledge. This requires the ability to speak clearly and to converse with people in various settings, and to translate engineering and scientific information into easy-to-understand concepts.

Writing skills: Civil engineers must be able to communicate with others, such as architects, landscape architects, urban and regional planners. They also must be able to explain projects to elected officials and citizens. Civil engineers must be able to write reports that are clear, concise, and understandable to those with little or no technical or scientific background.

Geotechnical Engineering

From a scientific perspective, geotechnical engineering largely involves defining the soil's strength and deformation properties. Clay, silt, sand, rock and snow are important materials in geotechnics. Geotechnical engineering includes specialist fields such as soil and rock mechanics, geophysics, hydrogeology and associated disciplines such as geology. Geotechnical engineering and engineering geology are a branch of civil engineering.

The specialism involves using scientific methods and principles of engineering to collect and interpret the physical properties of the ground for use in building and construction. Its practical application, e.g. foundation engineering, has come to require a scientific approach. The term geotechnics is currently used to describe both the theoretical and practical application of the discipline.

Geotechnics is applied when planning infrastructure such as roads and tunnels as well as buildings and other constructions onshore and offshore. The discipline also involves performing numerical calculations, analyzing the stability of slopes and cliffs, and assessing load-bearing capacity, settlement and deformation in man-made structures.

Research and development in geotechnical engineering is carried out to improve and further refine equipment and methods for carrying out ground surveys,

- Equipment and methods for surveying and testing sediment and rock samples in a laboratory,

- Methods for calculating and analysing the behaviour and bearing capacity of soil and rock when planning structures (buildings, bridges, dams etc.), offshore installations, tunnels and subterranean spaces, roads, railways etc.,

- Methods for measuring, instrumenting and subsequently documenting whether buildings and other structures behave the way they were designed to.

Geotechnical Engineer

Geotechnical engineers figure out the impact that geological formations may have on construction projects. They use advanced knowledge of scientific and mathematical processes to examine the formation of the earth beneath and around residential, commercial or industrial buildings and structures.

A geotechnical engineer's skills are used for drilling wells, constructing production and storage facilities, transporting petroleum products and examining ground water flow. This career has amazing possibilities, from marine operations, to floating ice platforms in the Arctic, to mining operations.

Work of a Geotechnical Engineer

All construction takes place in or on the ground, so it is easy to see how geotechnical engineering plays a crucial role in all civil engineering projects. Before any construction work takes place, it is vitally important to do a site investigation. Failure to carry this out often has had negative and expensive consequences on construction projects.

Geotechnical engineers guard and maintain the earth's physical environment during the development of major public and private projects. Combining their expertise in civil engineering construction and design enables them to safely investigate and analyze sites and determine their present and future stability. Projects like these typically involve major changes to the physical environment, and can include tunnelling and construction of major structures like buildings, bridges, dams, airport runways, and towers.

Geotechnical engineers perform the following functions within the framework of the following jobs:

Geotechnical or Geological Engineers (General)

Provide analysis and mapping of technical results obtained from seismic surveys, and investigate subsurface conditions and materials to determine their properties and risks.

Geotechnical or Geological Engineer (Oil Sands Projects)

Design open pit walls, mine waste dumps and dam structures used in oil sands mining, and analyze slope stability, seepage and hydraulic separation on dam structures.

Hydrogeological Engineer

Provide design and analysis of ponds containing discarded oil sands materials, water extraction from soil and sand, and steam injection into wells; and evaluate underground water layers trapped in rocks (aquifers). They also provide advice on environmental restoration.

Reservoir Geomechanics Engineer (Oil & Gas Operations)

Analyze the strength of soils, drill hole stability, stress constraint, permeability of rock formations and the degree of trapped hydrocarbons in underground reservoirs

Geomechanics Engineers (Marine Operations)

Analyze the relationship between physical structures and marine geology, anchoring systems, sediment erosion, slope stability, and foundations for offshore and coastal structures

Workplace of a Geotechnical Engineer

Geotechnical engineers spend most of their time working in comfortable office settings. They occasionally visit operation sites, and are sometimes exposed to potentially hazardous conditions and inclement weather. Extended visits do occur and on occasion, relocation may be required.

Geotechnical engineers can be employed by the following types of organizations:

- Colleges and universities
- Construction contractors
- Electrical utility companies
- Engineering consulting firms
- Mining companies
- Municipal, regional and federal governments
- Oil and gas exploration, production and transportation companies
- Petroleum services companies

- Public and private research organizations
- Real estate development companies.

Geomechanics

Geomechanics is the study of how soils and rocks deform, sometimes to failure, in response to changes of stress, pressure, temperature and other environmental parameters. In the petroleum industry, geomechanics tends to focus on rocks, but the distinction becomes blurred because unconsolidated rocks can behave like soils.

Geomechanics is relatively young as a science and even younger in its application to the petroleum industry. However, it applies to nearly all aspects of petroleum extraction from exploration to production to abandonment and across all scales, from as small as the action of individual cutters on a polycrystalline diamond compact (PDC) bit through drilling wells and perforating to as large as modeling fields and basins. Over the last 30 years, geomechanics has come to play an increasingly important role in drilling, completion and production operations. This trend continues as operators pursue oil and gas production from shales, in which mechanical anisotropy—the variation of mechanical properties with orientation plays a vital role.

At the wellbore scale, geomechanics is central to understanding how drill bits remove rock, characterizing borehole stability, predicting the stability of perforation tunnels and designing and monitoring hydraulic fracturing stimulation programs. At the reservoir scale, geomechanics helps model fluid movement and predict how fluid removal or injection leads to changes in permeability, fluid pressure and in situ rock stresses that can have significant effects on reservoir performance. Engineers use geomechanical modeling to predict and quantify these effects for life-of-reservoir decisions such as placing and completing new wells, enhancing and sustaining production, minimizing risk and making new investments.

Choosing the correct bit type and design for optimal rate of penetration and bit life is vital for drilling cost-effective wells. The geomechanics of rock destruction under the drill bit is complex because of high strain rates and temperatures, multiple deformation mechanisms and interactions between the bit, drilling fluid and formation. Many ad hoc approaches to understanding and improving drillbit performance have been taken, and interpretation methods such as mechanical specific energy the energy used to remove a unit volume of rock have been used since the 1960s to relate drilling performance to rock strength. Recent advances in research methods related to geomechanics are starting to reveal in more detail the factors, such as the balance between crushing, disaggregation and brittle cracking or chipping, that control the loads on the cutter, its wear behavior and the nature of the rock debris.

Geomechanics also plays a major role in understanding the stability and integrity of the borehole while drilling. The presence of the wellbore and the pressure of the drilling fluid induce changes in the stress state or in the rock. As a result, the rock around the borehole may fail if the redistributed stresses are greater than the rock strength. Tensile cracking occurs if the mud pressure becomes too high and causes the borehole wall to go into tension. Breakouts, which occur if the mud pressure becomes too low, are regions of enlargement on opposing sides of a borehole, where shear cracking has occurred and the broken rock has been removed by the bit, stabilizers or mud flow. These failures can lead to stuck pipe, lost circulation and other drilling problems but can also be a valuable source of information about stress magnitudes and orientations.

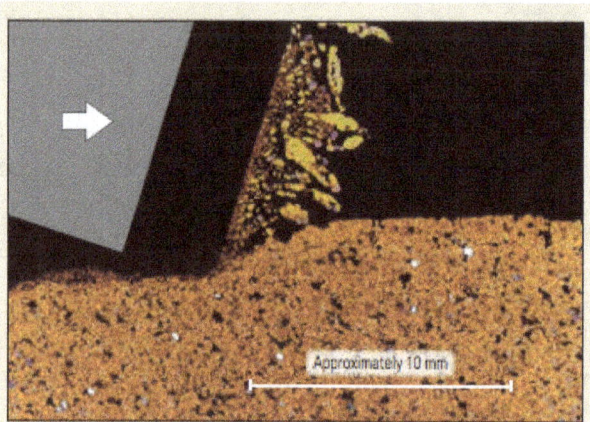

Cutting rock.

The action of a PDC cutter plowing through sandstone (orange) is shown in a cross section that was imaged using micro CT scanning. The location and direction of movement of the tungsten carbide (gray) and diamond (black) cutter are shown schematically. The section displays the deformation—crushing, disaggregation and brittle cracking—induced by the cutter.

By computing the stress changes around a planned borehole and comparing them to the strength of the rock, engineers can generate a mud weight program for the well. Typically, the mud weight must be high enough to suppress shear failure and fluid influx and low enough to avoid tensile failure and lost circulation. Although the mathematical techniques for calculating the stresses and the failure criteria are established and routine, the links between rock failure and drilling problems are not so well established. Furthermore, some additional failure modes, such as bedding plane failure, cannot yet be reliably predicted. Consequently, effective wellbore stability control, especially along challenging well trajectories, needs real-time monitoring of wellbore conditions as well as predrill prediction of the required mud pressure program.

A challenge for geomechanics modeling and prediction is the availability of input data—primarily rock strength and in situ stresses. Rock strength is easily measured on core samples in the laboratory, but the process is time-consuming and costly, and the results usually help with future wells rather than the current one. Consequently, considerable effort is spent deriving rock strength values from wireline, LWD and sonic data. The trade-off is less accuracy but higher spatial coverage along the well than is available from core data.

These data are interpolated or extrapolated to cover sections of interest in new wells or are used to improve predictions for the current well. The same data may also be used in geologic burial history models for constructing vertical profiles of the in situ stresses, which then are compared with and calibrated to discrete stress measurements in a well. More recently, advanced sonic tools have allowed estimation of rock strength and some components of the in situ stress for input into geomechanics models.

Events such as induced seismicity in the 1970s at the Rangely field, Colorado, USA, and compaction and subsidence in the 1980s at the Ekofisk field, North Sea, offshore Norway, helped engineers realize the role of geomechanics at the reservoir scale. Examining geomechanical changes on this scale is routine now, thanks to the development of finite element analysis programs that have been optimized for geologic structures and rock mechanical behavior. Populating these models with rock data can be a challenge, but since the computation grid is coarse, it can be done using seismic data. Once the model is populated, the mechanical response of the reservoir and overburden can be estimated for a variety of operations—including production, injection and fracturing. The model can be calibrated or refined with repeat, time-lapse seismic surveys and the addition of data as new wells are drilled. Operators can use this type of information to estimate the injection pressure used in fracture stimulations that risk a breach in a reservoir seal, or they can predict the fracture gradient after a period of production, allowing safe and effective drilling of infill wells.

Reservoir stimulation by fracturing, one of the first applications in the oil field to use geomechanics methods, is still a major development area. Exploitation of shale reservoirs has caused a surge of interest in the mechanical anisotropy of rock, which was not widely appreciated until about 2000. To make improved predictions of fracture geometry and growth, models for stress and strength and interpretations of sonic and resistivity measurements must be modified to account for anisotropy. Advances in sonic logging tools and interpretation have made this possible.

One feature common to all of these areas is the mechanical earth model (MEM), which is a collection of the data needed to make quantitative and qualitative predictions of the subsurface geomechanical environment. These data include the stresses in the Earth, pore pressure, rock elastic properties, strength and fabric and nonnumerical data such as the presence of intense natural fracturing. An MEM can be simple or complex, be large or small and be 1D, 2D, 3D or 4D—three spatial dimensions plus time— according to the complexity of the field and phenomena of interest. The most important defining feature of an MEM is that its data are related to the rocks that are being drilled, fractured or otherwise affected by field operations, rather than a particular well or set of wells. A second feature is that it is designed to be updated as new data become available from ongoing operations. Data sources for an MEM include any that give information on stress and mechanical behavior; such sources include wireline and LWD logs, cores, cavings and cuttings, regional geology and all types of seismicity.

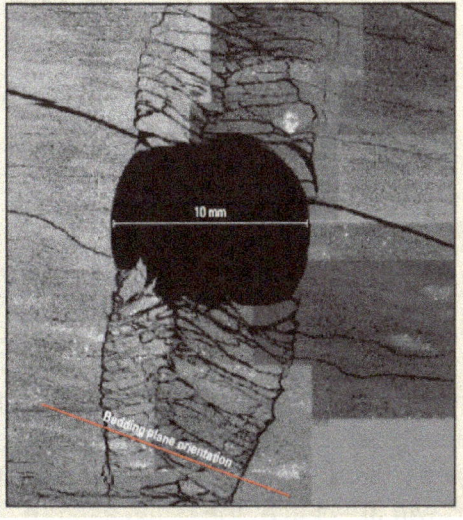

Unpredictable wellbore failure.

A laboratory model shows bedding plane failure in a hole drilled parallel to bedding in a fissile shale. The stress applied to the sample is the same in all directions, in spite of the directionality of the failure. This geometry is similar to the geometry of failures observed in the roofs of mines in fissile rocks, although the driving force in mines is gravity rather than in situ stress.

Coupled simulation.

A geomechanics model of a reservoir is shown with the seismic dataset (left) used to build it. Data for refinement of the model were acquired from several wells (colored lines), which were used to model changes over time from events such as depletion. Several faults (colored planes) and the top of the reservoir unit of interest are shown. The edges are the model boundaries or are reservoir discontinuities caused by faults.

Ongoing challenges for geomechanics include improvements to:

- Sources of data for predicting in situ stress and rock properties

- The use of anisotropic information for predicting deformation during exploitation of unconventional resources

- The treatment of fault and fracture displacements within numerical models.

In addition, to help improve the application of geomechanics to various sectors in the industry, engineers must have a better understanding of the relationships between rock failure and operational failure for wellbore instability and sand production.

Geomechanics in the oil field has come a long way from its early days as an adjunct to sonic logging. It is recognized as an important part of nearly all aspects of petroleum extraction and has been crucial in improving efficiency and driving down costs. The application of geomechanics in new reservoir types and mature ones and its integration into operators' workflows, along with the introduction of new measurements and techniques, will ensure its continuing role in the industry. From here, its operational impact will only grow. The application of geomechanics for revitalizing mature fields is imperative and will affect activities such as infill drilling, compaction mitigation and refracturing.

References

- Foundation-in-construction-purpose-functions-18963: theconstructor.org, Retrieved 28 May 2018

- Civil-engineers, careers: collegegrad.com, Retrieved 09 July 2018

- Foundation-engineering-of-structures: skyfilabs.com, Retrieved 11 July 2018

- Civil-Engineering, Engineering-branches: myklassroom.com, Retrieved 25 April 2018

- What-is-Geotechnical-engineering, Careers: ngi.no, Retrieved 11 May 2018

- Geotechnical-engineer, careers: sokanu.com, Retrieved 26 June 2018

2

Types of Foundations

A foundation is an architectural element which connects a structure to the ground. This leads to the easy transference of load. There are various kinds of shallow and deep foundations. This chapter closely examines such types of foundations, such as rubble trench foundation, shoring, underpinning and monopole foundation, among others.

Shallow Foundation

Shallow foundations are also called spread footings or open footings. The 'open' refers to the fact that the foundations are made by first excavating all the earth till the bottom of the footing, and then constructing the footing. During the early stages of work, the entire footing is visible to the eye, and is therefore called an open foundation. The idea is that each footing takes the concentrated load of the column and spreads it out over a large area, so that the actual weight on the soil does not exceed the safe bearing capacity of the soil.

In cold climates, shallow foundations must be protected from freezing. This is because water in the soil around the foundation can freeze and expand, thereby damaging the foundation. These foundations should be built below the frost line, which is the level in the ground above which freezing occurs. If they cannot be built below the frost line, they should be protected by insulation: normally a little heat from the building will permeate into the soil and prevent freezing.

Shallow foundations are used when the soil has sufficient strength within a short depth below the ground level. They need sufficient plan area to transfer the heavy loads to the base soil. These heavy loads are sustained by the reinforced concrete columns or walls (either of bricks or reinforced concrete) of much less areas of cross-section due to high strength of bricks or reinforced concrete when compared to that of soil. The strength of the soil, expressed as the safe bearing capacity of the soil is normally supplied by the geotechnical experts to the structural engineer. Shallow foundations are also designated as footings. The different types of shallow foundations or footings are discussed below:

1. Plain concrete pedestal footings

Plain concrete pedestal footings are very economical for columns of small loads or pedestals without any longitudinal tension steel. In figure above, the angle α between the plane passing through the bottom edge of the pedestal and the corresponding junction edge of the column with pedestal and the horizontal plane shall be determined.

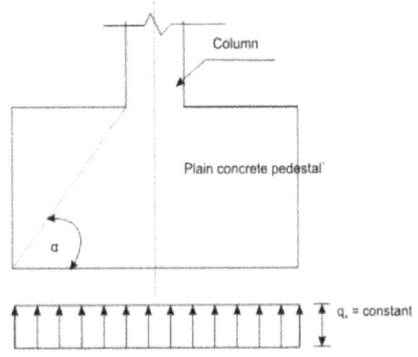

Plain concrete pedestal

2. Isolated footings

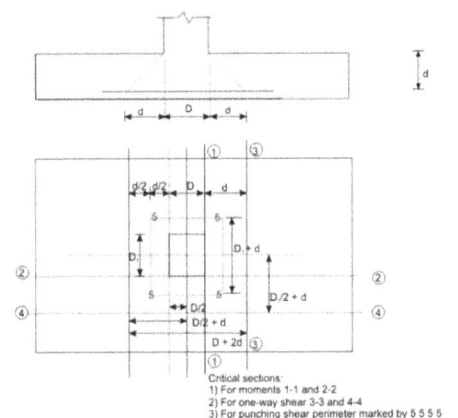

Uniform and rectangular footing

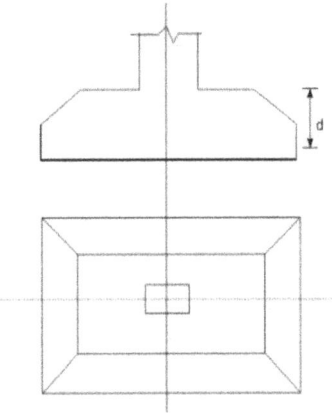

Sloped and rectangular footing

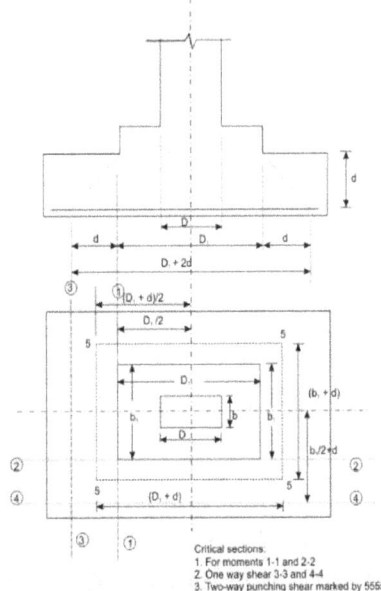

Stepped and rectangular footing

These footings are for individual columns having the same plan forms of square, rectangular or circular as that of the column, preferably maintaining the proportions and symmetry so that the resultants of the applied forces and reactions coincide. These footings, consist of a slab of uniform thickness, stepped or sloped. Though sloped footings are economical in respect of the material, the additional cost of formwork does not offset the cost of the saved material. Therefore, stepped footings are more economical than the sloped ones. The adjoining soil below footings generates upward pressure which bends the slab due to cantilever action. Hence, adequate tensile reinforcement should be provided at the bottom of the slab (tension face). the sloped or stepped footings, designed as a unit, should be constructed to ensure the integrated action. Moreover, the effective cross-section in compression of sloped and stepped footings shall be limited by the area above the neutral plane. Though symmetrical footings are desirable, sometimes situation compels for unsymmetrical isolated footings (Eccentric footings or footings with cut outs) either about one or both the axes.

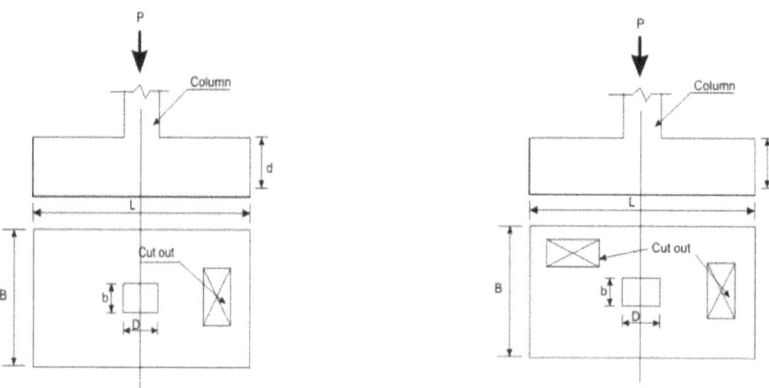

Figure: Unsymmetrical footing about x axis   Figure: Unsymmetrical footing about both axes

3. Combined footings

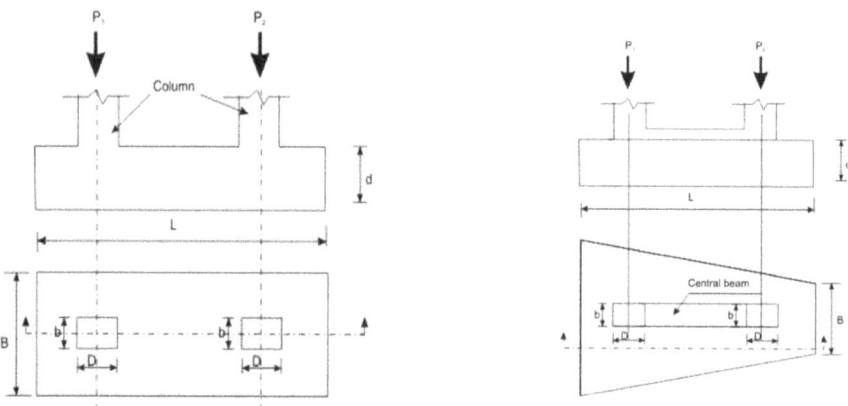

Figure: Combined footing without a central beam   Figure: Combined footing with a central beam

When the spacing of the adjacent columns is so close that separate isolated footings are not possible due to the overlapping areas of the footings or inadequate clear space between the two areas of the footings, combined footings are the solution combining two or more columns. Combined footing normally means a footing combining two columns. Such footings are either rectangular or trapezoidal in plan forms with or without a beam joining the two columns.

4. Strap footings

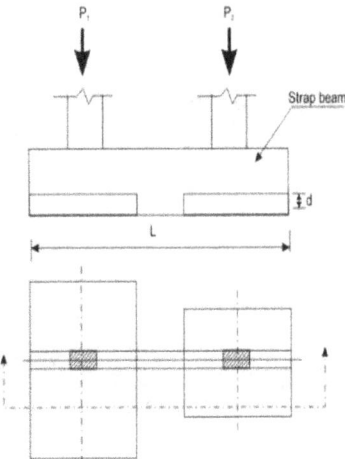

Figure: Strap footing

When two isolated footings are combined by a beam with a view to sharing the loads of both the columns by the footings, the footing is known as strap footing. The connecting beam is designated as strap beam. These footings are required if the loads are heavy on columns and the areas of foundation are not overlapping with each other.

5. Strip foundation or wall footings

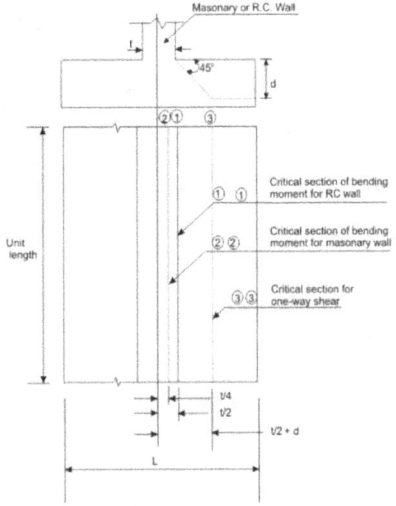

Figure: Wall footing

These are special cases of combined footing where all the columns of the building are having a common foundation. Normally, for buildings with heavy loads or when the soil condition is poor, raft foundations are very much useful to control differential settlement and transfer the loads not exceeding the bearing capacity of the soil due to integral action of the raft foundation. This is a threshold situation for shallow footing beyond which deep foundations have to be adopted.

Rubble Trench Foundations

Various forms of the rubble trench foundation have been used for thousands of years in construction. Earthen walls in the Middle East and Africa, for example, are built on top of shallow ditches filled with loose rock. Frank Lloyd Wright came across the rubble trench foundation system around the turn of the 20th Century. He observed the structures to be "perfectly static" with no signs of heaving, and thereafter built consistently with what he termed the "dry wall footing". Many time-tested structures stand as testimony to the longevity of the rubble trench.

Drainage and Foundation Combined

The rubble trench is a favourite type of foundation among many natural builders due to the fact that it substantially reduces cost, time, labor and the need for cement and rebar. It allows you to even eliminate the need for cement and rebar completely if needed, It also combines two key features that any successful foundation needs to achieve drainage and an even load distribution.

The Frozen Landscape

In temperate climates where winter frost should be expected a foundation must be designed to keep water from accumulating underneath the foundation walls in order to avoid frost heaving. Frost heaving is when water in the ground expands as it freezes and subsequently pushes the ground upward. And in spring the water thaws and the ground sinks back down again. This constant pushing up and down will force the house to twist and skew which might not only create problems with opening and closing of doors and windows but even structurally compromise the house.

The RTF deals with these challenges by consisting mainly of a drainage trench, much like a French drain, filled with crushed stone that lets any water drain down and away from the foundation. The basic principle of this has been used in many shapes and forms for thousands of years but it was American architect Frank Lloyd Wright who brought this technique (which he called "dry wall footing") into the 20th century.

The Trench

The trench should run underneath all the external load bearing points, continuously around the full perimeter of the building. Any internal load bearing points (underneath the house) can simply rest on undisturbed earth (stable soil cleared of top soil) so the amount of digging needed is minimal compared to your standard concrete slab.

The minimum depth of the trench is determined by the local frost-free depth. This is the depth where the temperature never descends to zero °C so any water ending up in the trench will therefore not freeze. The bottom of the trench then needs to slope with an even descent of at least 3 cm for every 1 meter of trench, diverting the water towards one point. From then on the water can be diverted away from the foundation through an outlet to either daylight or to a dry well.

Filter Fabric Geotextile

The trench must then be lined with a geotextile to prevent any of the surrounding soil from clogging up the trench and outlet. This geotextile is really the only non-natural material that you need for this foundation to work properly. The trench would probably work quite well without it for at least a while but any silting would gradually reduce the trench's ability to drain away water, eventually rendering the rubble trench completely useless, with frost heaving just waiting to happen.

Crushed Stone

After the trench is lined with Geotextile it is gradually filled with angular and washed stones that have an average size of between 2,5-5 cm, compacted at every 30 cm layer using either a hand powered tamper or a pneumatic tamper. The crushed stone needs to be washed before filling the trench otherwise it will contain sand and smaller particles that could silt up the trench and/or outlet over time. You continue filling the trench with crushed stone and tamping it every 30 cm layer until you reach about 20-30 cm below grade (ground surface). This is where the stonework begins.

Stonework

At this point you could of course construct a formwork, setup with rebar inside it and pour a concrete grade beam, in which case the rubble trench would greatly reduce the need for cement. However, the RTF can be constructed without a single drop of cement, using the oldest building material known to man; stone. It needs to be a continuous wall made of either really large blocks of stone, large enough for one row to sufficiently lift the house above ground or smaller, more easily managed stones arranged in a beautiful dry stacked stone wall. Both choices would create a look that definitely matches the natural materials of your house a lot better than cement ever could, and lets the building truly marry the landscape. The foundation should extend at least 40 cm above grade to keep the house away from any possible splashing and normal masses of snow.

First Line of Defense

The rubble trench should now be able to deal with most of the percipitation that one can expect but as an extra precaution you should add a small slope all around your entire foundation. This works as a first line of defense by keeping much of the precipitation from ending up in the trench in the first place. Simply let the Geotextile continue a bit up along the side of the foundation stones and fill up with earth against the stones with the Geotextile in between. This way you keep any of the added earth from ending up in the trench, avoiding any potential clogging. Then shape and compact the earth into a slope that leads water at least 1 meter away from the foundation. When done correctly this should leave you with a very dry and strong footing that will do its job for a very long time.

Benefits

Lower Cost: A rubble trench foundation requires less labor, uses less material, and reduced material costs compared to a standard concrete footing. There is no over dig, no footer forming, and no backfill.

Minimal Site Impact: Digging is limited to only the outline of the building, so site disruption is minimized.

Lower Greenhouse Gas Emissions. Rubble trench footers reduce concrete use by up to 80%, compared to a standard footer (depending on frost depth). Production of concrete requires a great deal of energy and generates 1.25 pounds of greenhouse gas for every pound of cement in the mix. Reducing total concrete used translates to direct reductions in greenhouse gas emissions.

Can Contain Recycled Content: The rubble fill can use recycled crushed concrete instead of gravel, as long as fine particles are washed out.

Improved Drainage: A rubble trench provides full water drainage under every structural bearing element of the foundation, ensuring that the footer remains dry at all times. This type of "static" foundation system ensures that no freeze-thaw heaving can occur.

Challenges

- Soils with low bearing capacity may require an extremely wide trench (or some other footing alternative) to achieve adequate bearing area.

- Rubble trench foundations are not specifically identified in building codes, so may require additional dialog with permitting officials. It helps to provide drawings stamped by a licensed engineer.

Deep Foundation

Deep foundation is required to carry loads from a structure through weak compressible soils or fills on to stronger and less compressible soils or rocks at depth, or for functional reasons. Deep foundations are founded too deeply below the finished ground surface for their base bearing capacity to be affected by surface conditions, this is usually at depths >3 m below finished ground level.

Deep foundation can be used to transfer the loading to deeper, more competent strata at depth if unsuitable soils are present near the surface.

Deep foundations are required when soil and superficial ground content is not stable or thick enough to support heavy loads. Deep foundations are achieved by forcing vertical structure components several feet below the ground's surface. The depth of the foundation is decided by how far stable soil lies beneath the ground surface.

Pile foundations are generally required to maintain safety in multi-story buildings. In addition, deep foundations in southern or coastal areas like Louisiana are necessary to gain stability despite a high water, silt, and loose clay ground content. In contrast, bedrock provides the most stable and safe building foundation. Pile foundations work to emulate a rock platform to provide a sturdy base.

Any raised structure will also require a pile foundation due to the concentration of building forces over a small area. The load bearing capacity of a pile depends on soil and structural capacity of the material from which the pile was made. Pile foundations also provide reinforcement against seismic activity and wind forces.

Types of Deep Foundation

Pile Foundation

Pile foundation is actually a slender column or long cylinder made of materials such as concrete or steel which are used to support the structure and transfer the load at desired depth either by end bearing or skin friction.

Pile foundations are deep foundations. They are formed by long, slender, columnar elements typically made from steel or reinforced concrete, or sometimes timber. A foundation is described as 'piled' when its depth is more than three times its breadth.

Pile foundations are usually used for large structures and in situations where the soil at shallow depth is not suitable to resist excessive settlement, resist uplift etc.

When to use Pile Foundation

Following are the situations when using pile foundation system can be:

- When groundwater table is high.
- Heavy and un-uniform loads from superstructure are imposed.
- Other types of foundations are costlier or not feasible.
- When the soil at a shallow depth is compressible.
- When there is the possibility of scouring, due to its location near river bed or sea shore etc.
- When there is a canal or deep drainage systems near the structure.
- When soil excavation is not possible up to the desired depth due to poor soil condition.
- When it becomes impossible to keep the foundation trenches dry be pumping or by any other measure due to heavy inflow of seepage.

Types of Pile Foundation

Pile foundations can be classified based on function, materials and installation process etc. Followings are the types of pile foundation used in construction:

Based on Function or Use

1. Sheet Piles
2. Load Bearing Piles
3. End bearing Piles
4. Friction Piles
5. Soil Compactor Piles

Based on Materials and Construction Method

1. Timber Piles
2. Concrete Piles
3. Steel Piles
4. Composite Piles

These piles are briefly discussed below.

Classification of Pile Foundation Based on Function or use

Sheet Piles

This type of pile is mostly used to provide lateral support. Usually, they resist lateral pressure from loose soil, the flow of water etc. They are usually used for cofferdams, trench sheeting, shore protection etc. They are not used for providing vertical support to the structure. They are usually used to serve the following purpose-

- Construction of retaining walls.

- Protection from river bank erosion.

- Retain the loose soil around foundation trenches.

- For isolation of foundation from adjacent soils.

- For confinement of soil and thus increase the bearing capacity of the soil.

Load Bearing Piles

This type of pile foundation is mainly used to transfer the vertical loads from the structure to the soil. These foundations transmit loads through soil with poor supporting property onto a layer which is capable of bearing the load. Depending on the mechanism of load transfer from pile to the soil, load-bearing piles can be further classified as flowed.

End Bearing Piles

In this type of pile, the loads pass through the lower tip of the pile. The bottom end of the pile rests on a strong layer of soil or rock. Usually, the pile rests at a transition layer of a weak and strong slayer. As a result, the pile acts as a column and safely transfers the load to the strong layer.

The total capacity of end bearing pile can be calculated by multiplying the area of the tip of the pile and the bearing capacity of at that particular depth of soil at which the pile rests. Considering a reasonable factor of safety, the diameter of the pile is calculated.

Friction Pile

Friction pile transfers the load from the structure to the soil by the frictional force between the surface of the pile and the soil surrounding the pile such as stiff clay, sandy soil etc. Friction can be developed for the entire length of the pile or a definite length of the pile, depending on the strata of the soil. In friction pile, generally, the entire surface of the pile works to transfer the loads from the structure to the soil.

The surface area of the pile multiplied by the safe friction force developed per unit area determines the capacity of the pile.

While designing skin friction pile, the skin friction to be developed at pile surface should be sincerely evaluated and a reasonable factor of safety should be considered. Besides this one can increase the pile diameter, depth, number of piles and make pile surface rough to increase the capacity of friction pile.

Soil Compactor Piles

Sometimes piles are driven at placed closed intervals to increase the bearing capacity of soil by compacting.

Classification of Piles Based on Materials and Construction Method

Primarily piles can be classified into two parts. Displacement piles and Non-displacement or Replacement piles. Piles which causes the soil to be displaced vertically and radially as they are driven to the ground is known as Displacement piles. In case of Replacement piles, the ground is bored and soil is removed and then the resulting hole is either filled with concrete or a pre-cast concrete pile is inserted. On the basis of materials of pile construction and their installation process load-bearing piles can be classified as follows:

1. Timber Piles

   i. Untreated

   ii. Treated with Preservative

2. Concrete Piles

   i. Pre-cast Piles

   ii. Cast-in-place Piles

3. Steel Piles

   i. I-Section Piles

   ii. Hollow Piles

4. Composite Piles

Timber Piles

Timber piles are placed under water level. They last for approximately about 30 years. They can be rectangular or circular in shape. Their diameter or size can vary from 12 to 16 inches. The length of the pile is usually 20 times of the top width.

They are usually designed for 15 to 20 tons. Additional strength can be obtained by bolting fish plates to the side of the piles.

Advantages of Timber Piles-

- Timber piles of regular size are available.
- Economical.
- Easy to install.
- Low possibility of damage.
- Timber piles can be cut off at any desired length after they are installed.
- If necessary, timber piles can be easily pulled out.

Disadvantages of Timber Piles-

- Piles of longer lengths are not always available.

- It is difficult to obtain straight piles if the length is short.

- It is difficult to drive pile if the soil strata are very hard.

- Spicing of timber pile is difficult.

- Timber or wooden piles are not suitable to be used as end-bearing piles.

- For durability of timber piles, special measures have to be taken. For example- wooden piles are often treated with preservative.

Concrete Piles

Pre-cast Concrete Pile

This type of piles is cast in pile bed in horizontal form if they are rectangular in shape. Usually, circular piles are cast in vertical forms. Precast piles are usually reinforced with steel to prevent breakage during its mobilization from casting bed to the location of the foundation. After the piles are cast, curing has to be performed as per specification. Generally curing period for pre-cast piles is 21 to 28 days.

Advantages of Pre-cast Piles

- Provides high resistance to chemical and biological cracks.

- They are usually of high strength.

- To facilitate driving, a pipe may be installed along the center of the pile.

- If the piles are cast and ready to be driven before the installation phase is due, it can increase the pace of work.

- The confinement of the reinforcement can be ensured.

- Quality of the pile can be controlled.

- If any fault is identified, it can be replaced before driving.

- Pre-cast piles can be driven under the water.

- The piles can be loaded immediately after it is driven up to the required length.

Disadvantages of Pre-cast Piles

- Once the length of pile is decided, it is difficult to increase or decrease the length of the pile afterward.

- They are difficult to mobilize.

- Needs heavy and expensive equipment to drive.

- As they are not available for readymade purchase, it can cause a delay of the project.

- There is a possibility of breakage or damage during handling and driving od piles.

Cast-in-Palace Concrete Piles

This type of pile is constructed by boring of soil up to the desired depth and then, depositing freshly mixed concrete in that place and letting it cure there. This type of pile is constructed either by driving a metallic shell to the ground and filling it with concrete and leave the shell with the concrete or the shell is pulled out while concrete is poured.

Advantages of Cast-in-Place Concrete Piles

- The shells are light weighted, so they are easy to handle.
- Length of piles can be varied easily.
- The shells may be assembled at sight.
- No excess enforcement is required only to prevent damage from handling.
- No possibility of breaking during installation.
- Additional piles can be provided easily if required.

Disadvantages of Cast-in-Place Concrete Piles

- Installation requires careful supervision and quality control.
- Needs sufficient place on site for storage of the materials used for construction.
- It is difficult to construct cast in situ piles where the underground water flow is heavy.
- Bottom of the pile may not be symmetrical.
- If the pile is un-reinforced and uncased, the pile can fail in tension if there acts and uplifting force.

Steel Piles

Steel piles may be of I-section or hollow pipe. They are filled with concrete. The size may vary from 10 inches to 24 inches in diameter and thickness is usually ¾ inches. Because of the small sectional area, the piles are easy to drive. They are mostly used as end-bearing piles.

Advantages of Steel Piles

- They are easy to install.
- They can reach a greater depth comparing to any other type of pile.
- Can penetrate through the hard layer of soil due to the less cross-sectional area.
- It is easy to splice steel piles.
- Can carry heavy loads.

Disadvantage of Steel Piles

- Prone to corrosion.

- Has a possibility of deviating while driving.
- Comparatively expensive.

Factors Affecting Selection of Pile Foundation Type

Few factors influence the selection of a particular type of pile foundation. These factors are noted below:

- Type and loads from the superstructure.
- Properties of soil.
- The depth of the soil layer capable of supporting the piles.
- Variations in length of pile required.
- Availability of materials.
- Durability required.
- Available equipment for pile driving.
- Budget.
- The depth of water level and intensity of underground water flow.
- Types of surrounding structures.

Causes of Pile Foundation Failure

Pile foundations may fail due to different reasons. One must take proper precautions before designing pile foundations so that the possibility of such failure reduces. Causes of failure of pile foundation are given below:

- Load implied on the pile is greater than designed load.
- Defecting workmanship.
- Dislocation of reinforcement of pile.
- End bearing pile resting on soft strata.
- Faulty soil investigation.
- Selecting the wrong type of pile.
- Under-reinforcement of the pile.
- A decay of piles. (like attack of insects, corrosion etc.)
- Deformation of piles due to lateral loads.
- Incorrect assessment of pile capacity.
- Not considering lateral forces for designing of piles.

Pier Foundation

A *pier foundation* is a collection of large diameter cylindrical columns to support the superstructure and transfer large super-imposed loads to the firm strata below. It stood several feet above the ground. It is also known as "post foundation".

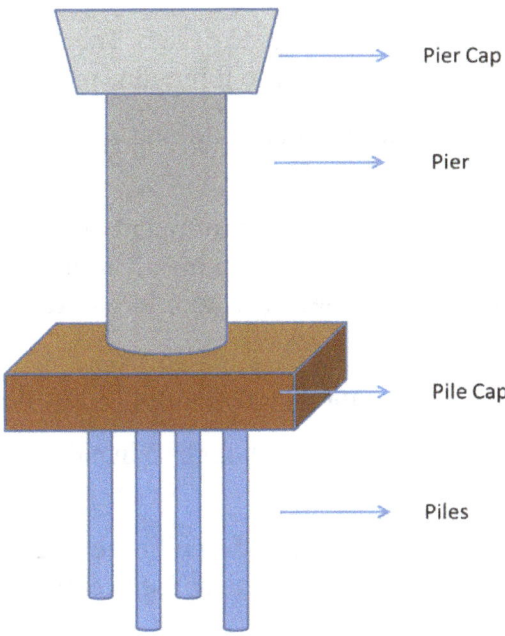

Types of Pier Foundation

Usually two types of pier foundation are used. These are:

1. Masonry or concrete pier

2. Drilled caissons

Masonry or Concrete Piers

Masonry or concrete piers depend on the level of the stratum. If a good bearing stratum exists up to 5 m, masonry piers are used. The size and the shape of the piers depend on the nature of the soil, depth of the bed etc.

Drilled Caissons

Drilled caissons usually refer to the cylindrical foundation. A drilled caisson is largely a compressed member subjected to an axial load at the top and reaction at the bottom. There are three types of drilled caissons:

•   Concrete caisson with enlarged bottom

•   Caisson of steel pipe with concrete filled in the pipe

•   Caissons with concrete and steel core in the steel pipe

Advantages of Pier Foundation

Pier Foundation is found in the coastal areas. There are many advantages of it:

- This method is easy and requires less amount of materials and labor. The materials required here is easily available

- It has a wide range of variety when comes to design. There are varied materials we can use here to increase the aesthetic view and also it remains in our budget

- Pier foundation save money and time as it doesn't need extensive excavation and lot of concrete

- It causes minimal disruption to the soil environment. a shovel can be used for excavation and existing roots and soil organisms remain mostly undisturbed. At the end of the building's useful life, the site will be easier to restore to a natural state than a site with a full basement.

- As it lifts the house above the ground, floods cannot do any damage to the structures.

- The space between house and ground is enough to install utilities like plumbing and electrical wires among them.

- Workers can easily get under the space between house and ground to solve issues related to plumbing and electrical as there is enough space to crawl

- It's comfortable to walk on the floor which doesn't rest on a firm surface and it is good for the people who have arthritis and back pain

- Inspection is possible as the diameter of the shafts are larger

- Engineers can change the design whenever they want if necessary

- The ground vibration that is normally associated with driven piles is absent in case of drilled pier construction.

- Bearing capacity can be increased by under-reaming the bottom (in non-caving materials).

Pier Foundation Construction Details

Spacing of Piers

Pier foundations usually are built 1-1.5 feet above the ground. This gap (how far apart are foundation piers) is necessary to prevent the moisture as the moisture damages the wooden structures.

Shape and Size of piers

The shape of the piers is:

- Square

- Rectangular

- Circle

The diameter of piers is usually 6 in, 8 in, 10in, 16in. The depth of pier foundation is below the freezing depth. The depth is around 5-6ft.

The piers can be made from varied materials

- Wood
- Brick
- Solid concrete

Ways of Performing

There are many ways to do pier foundation. Masonry is the most convenient way among them. But the process is not ideal. The stack of bricks is not directly put together in the hole. Fold pillar is completely on the ground, and then lower it into the wall and also does not look fast and enjoyable.

Materials used

Wood, Brick, concrete etc. are used in pier foundation. But the most used material is reinforced concrete. It provides most compressive strength and has a high tolerance tensile strain. Furthermore, reinforced monolithic pillars can withstand any type of frosting and will not crack under these forces. And it is quite easy to dilute the concrete mixture and pouring it into the holes.

Types of Cross Section

There are many types of foundation pillars' cross-sections. It can be cylindrical, helical or box-shaped or more complex form with the broadening of the bottom of the posts. The broadening can increase the area of the base and thus increase the load-bearing capacity of the foundation. The weight of the house will be distributed over a larger area.

Technologies of Pier Foundation

The most used way consists of digging a square or rectangular hole. The size of which is 4-8 inches longer than the required diameter of the post. Then formwork is set in the pit which settles the shape of the future foundation; then the reinforcement cage is placed, and concrete is poured. After that, the formwork is removed and covered with a pillar. This technology has made it possible for us to make a reinforced concrete pillar of different shapes but requires a relatively substantial number of earthworks and the use of removable shuttering. Installation of every new post for the foundation is the following - in the dug well we set formwork which propped up the sides of the spacer. Inside the formwork, there is the reinforcement cage. The reinforcement cage makes sure that the reinforcements remain in place.

Grillage

The top of the pier foundation which connects the individual posts into a single structure is called *grillage*. When heavy structural loads from columns, piers or stanchions are required to be transferred to a soil of low bearing capacity, grillage foundation is used. Grillage is often found to be lighter and economical.

Suitable Condition for Pier Foundation

Pier foundation is used in the below conditions:

- When decomposed rocks are present in the top strata, and there are underlying strata of sound rock below them, in such condition pier foundations, are used.

- As stiff clays offer a lot of resistance when driving a bearing pile, pier foundations can be conveniently used in such situation.

- It is used if the house is built from log, timber, frame as the pillars are small relative to other foundations.

- If structure needed to be built on a slope, pier foundation is used.

- The soil must have a low bearing capacity of water unless the pillars will sink under the weight of the house.

Caisson Foundation

In Civil Engineering, *Caissons* denote watertight structures which are constructed in connection with the excavation for foundations of bridges, piers, abutments in river and lake dock structure, foreshore protection etc. The caisson remains in its pose and thus ultimately becomes as integral parts of the permanent structure. It can be made up of wood, steel or reinforced concrete etc.

Requirement of Caisson

Followings are the suitable conditions for caisson foundation:

- When the soil contains large boulders, which obstruct penetration of piles.

- When a massive substructure is required to extend to or below the rear bed to provide resistance against destructive forces due to floating objects and score etc.

- When the foundation is subjected to a large lateral load.

- When the depth of water level in the river and sea is high.

- When there are river forces included in the load compositions.

- When the load is needed to carry at the end, caissons are preferred.
- When the present groundwater level is aggressive inflow, caissons are suitable.

Mechanism of Caisson

Caisson is a box but with no floor underneath it. So, when we put it underwater instead of filling up with water as it is airtight, bubbles form. So as a result, we have a dirt floor from where all the water is kept out. But water is heavy so the surface of the water is exerting pressure on the caisson and tries to enter the caisson. So, in order to solve this, we build a platform up in the top and a tube connecting the caisson to the platform that exerts compressed pressure into the surface of the caissons. The front pressure and the chamber pressure equalizes, therefore. The workers climb down in the box and they start digging the dirt out of it. Now if the dirt is taken out manually, all the water will come inside and drown everything. To solve this, we build a pipe full of water. Air pressure from the environment and inside pressure of the caisson keep the water intact in the pipe. Next, the workers send a bucket through the pipe and fill it with dirt and then carried back up. Thus, the workers can dig their way down the riverbed.

Types of Caisson

There are several types of caisson foundation:

- Box Caissons: Box caissons are concrete boxes that are put on pre-prepared bases. The top of these boxes is open. They are floated into the places and once placed in appropriate place, they are filled with concrete. They are used for the construction of bridge pier.
- Excavated Caissons: These kinds of caisson are used to excavate. Cylindrical shaped, these caissons are filled with concrete.
- Floating Caissons: These are prefabricated boxes filled with concrete. It is also called floating docks.
- Open Caissons: Open caissons are quite like box caissons. There are two types of open caissons. Top and bottom both open and open top-closed bottom. Soft soil is suitable for these kinds of caissons. Open Caissons are used in the formation of the pier, deep manholes, pump stations, micro-tunneling etc.
- Pneumatic Caissons: Watertight or Box caissons which are used in underwater construction are known as pneumatic caissons.
- Compressed Air Caissons: This type of caissons is suitable for parched working conditions where other methods might seem inconvenient.
- Monolithic Caissons: Larger size of caissons compared to others.

Construction Process

Following steps could be followed to build a caisson:

- We must set the place first for the establishment of the caisson.

- Then the first 3.7m of Caisson is pre-casted.

- Next, with the help of towboat, the caisson is floated to its location by and tie it to the caisson guide.

- After that, concrete is poured using slip forming and as concrete goes, the box becomes heavier and sinks into the water along the caisson guide.

- Mooring cables are used to hold the caissons in place.

- When the caisson finally touches river bottom, the mooring cables are removed.

- Finally, the cap is poured.

Uses of Caisson Foundation

Main functions of caisson foundation are given below:

- Caisson is used in building bridge piers as it stays in water almost all the time.

- Caisson is constructed in connection with excavation for the foundation of piers and abutments in rivers and lake, bridges breakwater dock structures for the point of view of shore protection, lamp house etc.

- It is also used for pump house which is subjected to huge vertical as well as horizontal forces.

- It is sometimes used for large and multi-storied buildings.

- Pneumatic caisson is used in railway bridges, garbage pits, water supply, sewage facilities etc.

- Caisson serves as an impervious core wall on earth dams when placed adjacent to each other.

- Caisson provides an access to the deep shaft or a tunnel.

- Caisson provides an enclosure below water level for installing machinery, pumps etc.

- Caissons have also been used in the installation of hydraulic elevators where a single stage ram is installed below the ground level.

Buoyancy Rafts Foundations

Buoyancy rafts or hollow box foundations also known as the floating foundations is a type of deep foundation is used in building construction on soft and weak soils.

The decision of construction of a deep foundation is taken when the load has to be distributed to an area either with soft weak compressible soils or to reach strata that have strong soil or rock or any other special concerns. One such generally used a type of deep foundation is called as Hollow box foundations or buoyancy rafts foundation.

These types of foundations are designed such a way that they behave as buoyant (floating) sub-structures for the net loading over it. Hence reducing the load intensity over the soil.

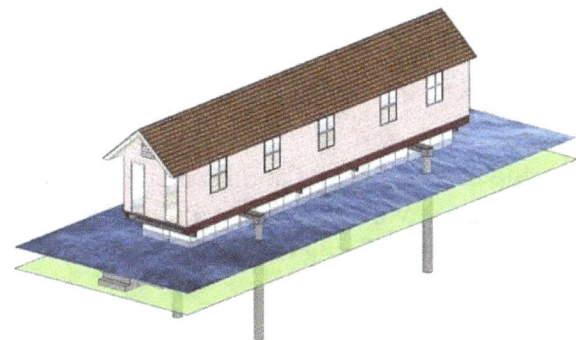

The buoyancy raft foundations are also known as compensated foundations or deep cellular rafts. Whatever be the name they are known for, they come under the category of floating foundations.

These are known as fully compensated foundations as during their construction the soil underneath the foundation is removed, whose weight is equal to the weight of the whole superstructure. Hence the weight removed from the soil is compensated by the weight of the building.

The buoyancy rafts are adopted under the following cases:

- The soil bearing capacity is very low

- The estimated building settlement is more than the safe limit

Need for Buoyancy Rafts in Building Construction

Those areas with soft soil, layer in a huge depth is observed, it is economical to go for floating foundation. No other foundation like pile foundation cannot be an efficient replacement for these.

Occupants in low-altitude areas face the problem of high floods resulting in the collapse of houses. The construction of buoyant foundation would help in increasing the elevation of the house.

This arrangement is a flexible method, as the building would remain on the ground under the normal conditions. When flood approaches, the building would rise to a necessary height, as shown in the figure.

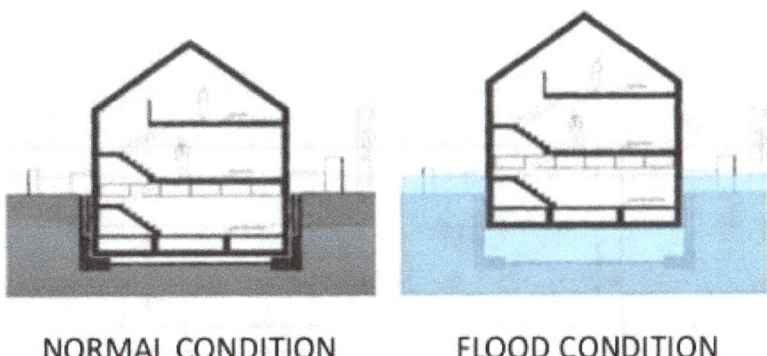

NORMAL CONDITION   FLOOD CONDITION

Figure: Buoyancy Rafts or Hollow Box Foundation in case of Flood

Types of Buoyancy Rafts or Hollow Box Foundations

Now the floating foundation can be of two types:

1. The basement rafts, and

2. The buoyant raft.

The buoyant rafts differ from the basement rafts and both should not be confused. The basement rafts, unlike buoyant raft foundation, involve only excavation of soil whose weight equals to a part of the weight of the building. The whole building weight is not taken and hence called as a partially compensated foundation.

Construction of Buoyancy Rafts or Hollow Box Foundations

In the case of buoyancy rafts construction, the shear strength of the soil in the site is very low. Now under such conditions, the construction by floating of the foundation is the only way that works.

To have such a foundation, cellular rafts are sunk in the form of the box section. These forms a raft foundation that is rigid in nature, which reduces the settlement.

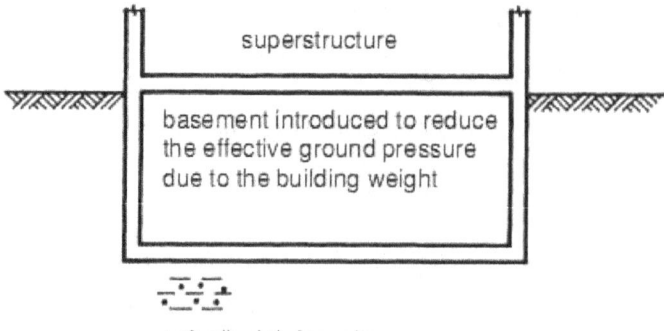

Figure: Arrangement of Buoyant Foundations

As shown in the figure above, the load from the soil is reduced and hence the superstructure floats like a boat. The bottom basement as explained is placed on the excavated area.

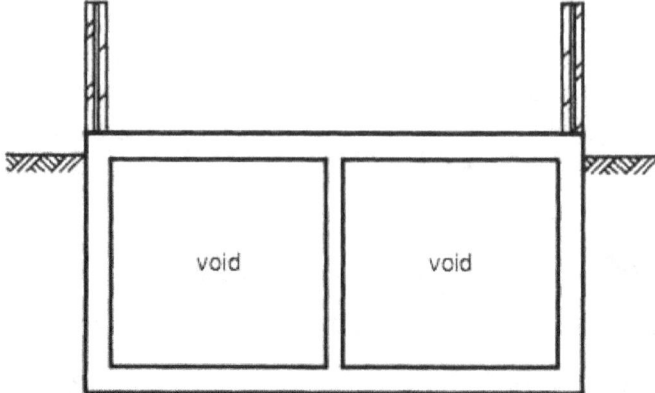

Figure: Arrangement of Cellular Buoyant Raft Foundation

As shown in the figure above, the bottom slab can be treated as the basement for the foundation. This is connected to the ground slab, which forms a raft foundation. The foundation can be made cellular as shown.

By the usage of hollow raft or cellular rafts substructure, the total load value contributed by the building and the foundation itself is reduced to a lower value by the soil that is excavated, which make us clear about the designation, fully compensated foundations.

Caisson Type Buoyancy Rafts

These forms cellular caissons as shown in the figure below. This method is found economical, except for the cases during the sinking. The sinking may cause disturbances to the nearby soil causing further loosening of the soil arrangement.

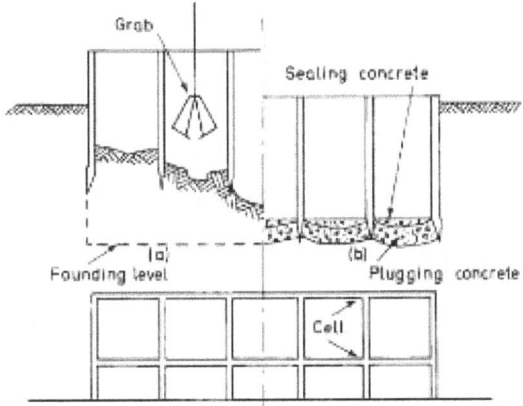

Figure: Buoyancy Rafts with Cellular Caissons

The construction of these rafts can be limited to certain individual areas or in the form of strips. Later these are connected to have a proper bond. This limited use would help in reducing the amount of excavation conducted in the construction area.

The figure below shows a construction of a buoyancy raft or hollow raft foundation for a G+15 Building in Glasgow.

Figure: Construction of a Buoyancy Raft or Hollow Raft
Foundation for a G+15 Building in Glasgow

Design Considerations for Buoyancy Raft or Hollow Raft Foundation

The general design involves following step by step procedures:

1. The initial step involved is to determine the depth of the excavation. Its size is determined by the plan of the building.

2. Next is to calculate the center of gravity that is required for the overburden removal that must be equal to sustain the structural buoyancy.

3. Compare the decision obtained in basement design with the client's needs as well as with his opinions.

4. Calculation of the water pressure, to check for the flotation phenomenon.

5. To bring up a design combining all the needs and requirements obtained from above four steps. This would give us avoided foundation.

6. The final step in design is to prepare to design details for external walls, floors as well as the separating walls. The main concern is to design for earth pressure, the bending moments and the shear forces. Special design for flotation must be considered.

Advantages of Buoyant Raft Foundation

- The building is elevated to a required height, which makes them stay above high water levels.

- The chances of settlement are reduced. As the total weight coming is equal to the excavated soil, there is no settlement. A slight increase in this load would not bring any drastic change in exceeding the settlement limit.

- Watertight material for foundations would help in the durability of the underground structure.

- Foundation having a height of 3 or 4 floors are used for skyscraper construction. This construction is found for efficient than pile foundation in areas with very weak soils for a larger depth.

Disadvantages of Buoyant Raft Foundation

- As the depth of weak soil increases, the amount of excavation also increases. In construction, the excavation process is an activity found very costly.

- Catastrophic Movement possibilities- These occur when the excavation is done beyond the critical depth of stability.

- Support of deep foundation may undergo settlement due to the earth pressure it is subjected to, from the surrounding soft soils.

- Space is wasted as hollow raft foundations cannot be used for any other purpose.

- Uplift pressure would affect the whole foundation arrangement.

- Leaving the cells unoccupied would result in water entering or seepage. This may be through the substructure or due to any complaints in the water pipelines going underground.

- There are possibilities of leakage of gasses into these hollow cells. As these situations are unaware, they remain unventilated. This would cause unexpected explosions, even if there are small chances of ignition anywhere around.

Basement Foundation

A basement, in literal terms, is an eight foot (or deeper) hole that ends in a concrete slab.

For a long time, basement walls were built with cinder blocks. As a result, they were prone to structural failures and leaks as they aged.

These days all are built with poured concrete walls, which have virtually eliminated most problems related to structural integrity and moisture permeation.

Basements are excellent for anchoring a property to the ground while extending the foundation below the area's frost line, which helps maintain the integrity of the foundation over time.

The finished footers of a basement foundation that's ready to have the walls framed in and poured.

Advantages of Basement Foundations

- More, Cheap Square Footage: Arguably the greatest advantage to a basement foundation is the additional square-footage gained and at a much lower cost per square foot than other parts of the home.

- Seasonal Living Space: Great for smaller-footprint homes, the addition of a finished basement creates energy-efficient living spaces that pair well with changing seasons, staying warm in winter and cool in summer.

- Easy-Access for Repairs: It goes without saying that it's easier (and cheaper) for technicians to make repairs to your home's utilities standing up rather than crawling in a crawlspace or digging into a slab.

- Storm Protection: For you, the residents, as well as your home. Basements make great shelters from the worst mother-nature can throw at you, while still providing a solid anchor for your home.

Disadvantages of Basement Foundations

- Increased Cost: A basement is understandably the most expensive foundation-type of the three mentioned here moreso if you choose to finish that space. Even then, that finished basement square footage will most likely be the least expensive in your entire home.

- Potential Flooding: Without a sump pump, you may end up with a flooded basement. To combat (and virtually eliminate the threat of) flooding, we recommend battery backup, generators or water-flow backups. In all cases, though, the best scenario to prevent potential flooding is ensuring a natural path for drainage.

- Lack of Natural Light: If you're converting your basement into a living space, and it's NOT a walkout, you might have to find creative ways of bringing some light into the space. Again, this isn't an issue with walkout basements.

A walkout basement just after floor was poured.

Shoring

Shoring is the construction of a temporary structure to support temporarily an unsafe structure. These support walls laterally. Shoring can be used when walls bulge out, when walls crack due to unequal settlement of foundation and repairs are to be carried out to the cracked wall, when an adjacent structure needs pulling down, when openings are to be newly made or enlarged in a wall.

Types of Shoring

- Raking shoring
- Flying shoring
- Dead shoring

Raking Shoring

In this method, inclined members known as rakers are used to give lateral supports to walls. A raking shore consists of the following components:

- Rakers or inclined member

- Wall plate
- Needles
- Cleats
- Bracing
- Sole plate

The following points are to be kept in view for the use of the raking shores:

- Rakers are to be inclined in the ground at 45°. However the angle may be between 45° and 75°.
- For tall buildings, the length of the raker can be reduced by introducing rider raker.
- Rakers should be properly braced at intervals.
- The size of the rakers is to be decided on the basis of anticipated thrust from the wall.
- The centre line of a raker and the wall should meet at floor level.
- Shoring may be spaced at 3 to 4.5 m spacing to cover longer length of the bar.
- The sole plate should be properly embedded into the ground on an inclination and should be of proper section and size.
- Wedges should not be used on sole plates since they are likely to give way under vibrations that are likely to occur.

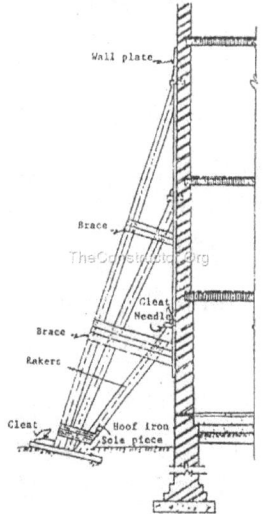

Raking Shores Wall Support

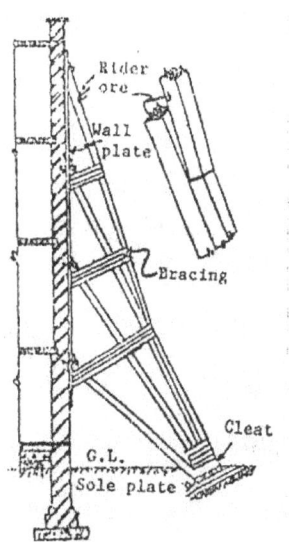

Raking shore for Multistoried Building where inclination of the rakers has to be limited due to short land width available

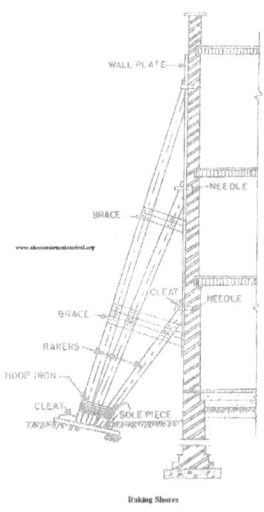

Detail of Head of the Raker Shores

Flying Shoring

Flying shores is a system of providing temporary supports to the party walls of the two buildings where the intermediate building is to be pulled down and rebuilt. All types of arrangements of supporting the unsafe structure in which the shores do not reach the ground come under this category.

The flying shore consists of wall plates, needles, cleats, horizontal struts (commonly known as horizontal shores) and inclined struts arranged in different forms which vary with the situation. In this system also the wall plates are placed against the wall and secured to it.

A horizontal strut is placed between the wall plates and is supported by a system of needle and cleats. The inclined struts are supported by the needle at their top and by straining pieces at their feet. The straining piece is also known as straining sill and is spiked to the horizontal shore. The width of straining piece is the same as that of the strut.

When the distance between the walls (to be strutted apart) is considerable, a horizontal shore cannot be safe and a trussed framework of members is necessary to perform the function of flying shore.

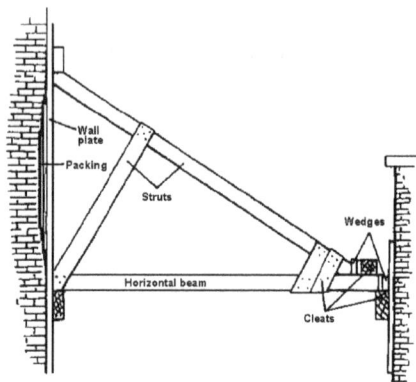

Flying Shore

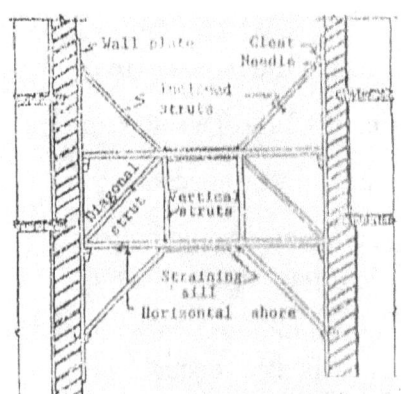

Flying shore when the distance between
two walls is considerable

Dead Shoring

Dead shore is the system of shoring which is used to render vertical support to walls and roofs, floors, etc. when the lower part of a wall has been removed for the purpose of providing an opening in the wall or to rebuild a defective load bearing wall in a structure.

The dead shore consists of an arrangement of beams and posts which are required to support the weight of the structure above and transfer same to the ground on firm foundation below.

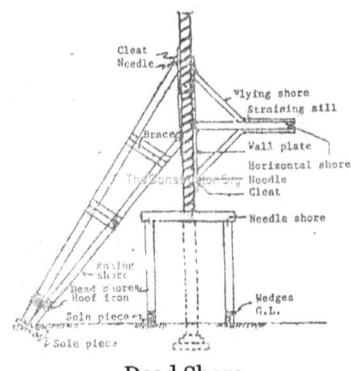

Dead Shore

When opening in the wall are to be made, holes are cut in the wall at such a height as to allow sufficient space for insertion of the beam or girder that will be provided permanently to carry the weight of the structure above.

Distance at which the holes are cut depends upon the type of masonry and it varies from 1.2m to 1.8m centre. Beams called needles are placed in the holes and are supported by vertical props called dead shores at their ends on either side of the wall. The needles may be of timber or steel and are of sufficient section to carry the load above.

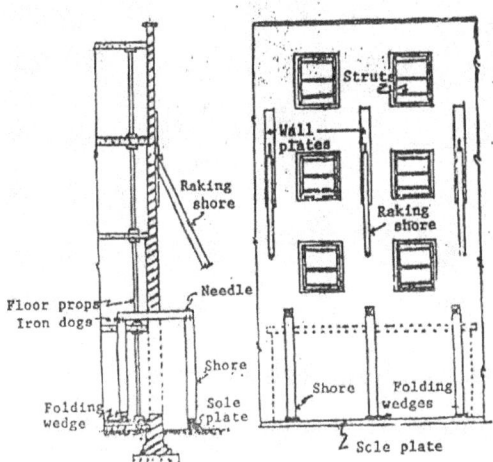

Section of the elevation showing arrangement of dead shores for making an opening in an existing wall

The dead shores stand away from wall on either side so as to allow for working space when the needle and the props are in position. The props are tightened up by folding wedges provided at their bases while the junction between the prop and the needle is secured with the help of dogs.

Before the dismantling work is started, all the doors, windows or other openings are well strutted. In order to relieve the wall of load of floors and roof above, they are independently supported.

Vibrations and shocks are bound to occur when wall cutting is done as such a measure of safety raking shores are sometimes erected before commencement of wall cutting operation.

Underpinning

Underpinning is a method for repair and strengthening of building foundations.

There are situations where a failure in foundation or footing happens unexpectedly after the completion of whole structure (both sub and superstructure). Under such an emergency situation, a remedial method has to be suggested to regain the structural stability.

The method of underpinning help to strengthen the foundation of an existing building or any other infrastructure. These involve installation of permanent or temporary support to an already held foundation so that additional depth and bearing capacity is achieved.

Underpinning Method

Selection of Underpinning Methods

Underpinning methods are selected based on age of structure and types of works involved.

Structure Categories Based on its Age

- Ancient Structures : Age greater than 150 years
- Recent Structures : Age between 50 – 150 years
- Modern Structure : Age less than 50 years

Types of Works for Selection of Underpinning Methods

Conversion Works

The structure has to be converted to another function, which requires stronger foundation compared to existing.

Protection Works

The following problems of a building has to undergo protection works:

- The existing foundation is not strong or stable
- Nearby excavation would affect the soil that supports existing footing.
- Stabilization of the foundation soil to resist against natural calamities
- Requirement of basement below an already existing structure

Remedial Works

- Mistakes in initial foundation design caused subsidence of the structure
- Work on present structure than building a new one

Structural Conditions which Requires Underpinning

There are many reasons that make an engineer to suggest underpinning method for stabilization of the substructure such as:

- The degradation of timber piles used as a foundation for normal buildings would cause settlement. This degradation of structures is due to water table fluctuations.

- Rise and lowering of the water table can cause a decrease of bearing capacity of soil making the structure to settle.

- Structures that are built over soil with a bearing capacity not suitable for the structure would cause settlement.

Need for Underpinning

The decision of underpinning requirement can be made based on observations. When an already existing structures start to show certain change through settlement or any kind of distress, it is necessary to establish vertical level readings as well as at the offset level, on a timely basis. The time period depends upon the how severe is the settlement.

Now, before the excavation for a new project, professionals have to closely examine and determine the soil capability to resist the structure that is coming over it. Based on that report the need for underpinning is decided. Sometimes such test would avoid underpinning to be done after the whole structure is constructed.

Methods of Underpinning

Following are the different underpinning methods used for foundation strengthening:

- Mass concrete underpinning method (pit method)
- Underpinning by cantilever needle beam method
- Pier and beam underpinning method
- Mini piled underpinning
- Pile method of underpinning
- Pre-test method of underpinning

Whatever be the types of underpinning method selected for strengthening the foundation, all of them follow a similar idea of extending the existing foundation either lengthwise or breadthwise and to be laid over a stronger soil stratum. This enables distribution of load over a greater area.

Different underpinning methods are mentioned briefly in the following sections. The choice of method depends on the ground conditions and the required foundation depth.

Mass Concrete Underpinning Method (Pit Method)

Mass concrete underpinning method is the traditional method of underpinning, as it has been followed by centuries. The method involves extending the old foundation till it reaches a stable stratum.

The soil below the existing foundation is excavated in a controlled manner through stages or pins. When strata suitable is reached, the excavation is filled with concrete and kept for curing, before next excavation starts.

In order to transfer the load from old foundation to new one, a new pin is provided by means of placing dry sand-cement pack. This is a low-cost method suitable for the shallow foundation.

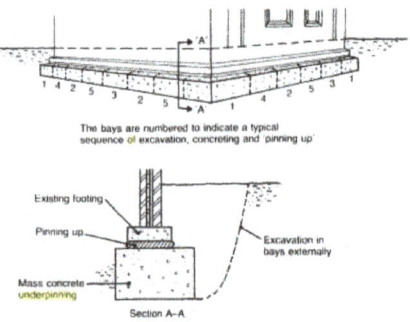

Mass Concrete Underpinning

For more complicated problems related to the foundation other superior methods have to chosen.

Underpinning by Cantilever Needle Beam Method

Figure below represents the arrangement of cantilever pit method of underpinning, which is an extension of pit method. If the foundation has to be extended only to one side and the plans possess a stronger interior column, this method can be used for underpinning.

Advantages of Cantilever Needle Beam Method

- Faster than traditional method
- One side access only
- High load carrying capability

Disadvantages

- Digging found uneconomical when existing foundation is deep
- Constraint in access restricts the use of needle beams

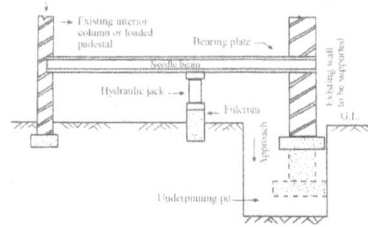

Cantilever Needle Beam Underpinning Method

Pier and Beam Underpinning Method

It is also termed as base and beam method which was implemented after the second world war. This method progressed because the mass concrete method couldn't work well for a huge depth of foundation.

It is found feasible for most of the ground conditions. Here reinforced concrete beams are placed to transfer the load to mass concrete bases or piers as shown in.

The size and depth of the beams are based on the ground conditions and applied loads. It is found economical for depth shallower than 6 m.

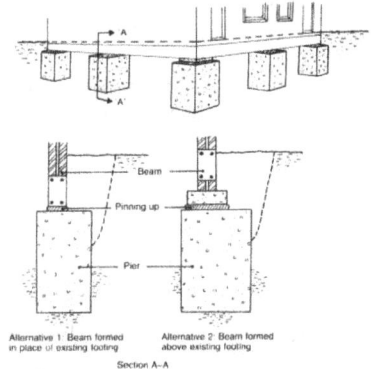

Pier and Beam Underpinning Method

Mini Piled Underpinning

This method can be implemented where the loads from the foundation have to transferred to strata located at a distance greater than 5m. This method is adaptable for soil that has variable nature, access is restrictive and causes environmental pollution problems.

Piles of diameter between 150 to 300 mm in diameter is driven which may be either augured or driven steel cased ones.

Pile Method of Underpinning

In this method, piles are driven on adjacent sides of the wall that supports the weak foundation. A needle or pin penetrates through the wall that is in turn connected to the piles as shown in figure below.

These needles behave like pile caps. Settlement in soil due to water clogging or clayey nature can be treated by this method

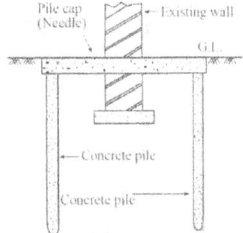

Underpinning by Pile Method

Pre-test Method of Underpinning

It is employed for strip or pad foundation. Can be used for building with 5 to 10 stories. Here the subsoil is made compact and compressed, in the new excavation level that gives predetermined loads to the soil. This is done before underpinning is performed.

Here reduced noise and disruption are expected. This method cannot be implemented for raft foundation.

Monopile Foundation

The mono pile foundation is a simple construction. The foundation consists of a steel pile with a diameter of between 3.5 and 4.5 metres. The pile is driven some 10 to 20 metres into the seabed depending on the type of underground. The mono pile foundation is effectively extending the turbine tower under water and into the seabed.

An important advantage of this foundation is that no preparations of the seabed are necessary. On the other hand, it requires heavy duty piling equipment, and the foundation type is not suitable for locations with many large boulders in the seabed. If a large boulder is encountered during piling, it is possible to drill down to the boulder and blast it with explosives.

Costs by Water Depth for Mono Pile Foundations

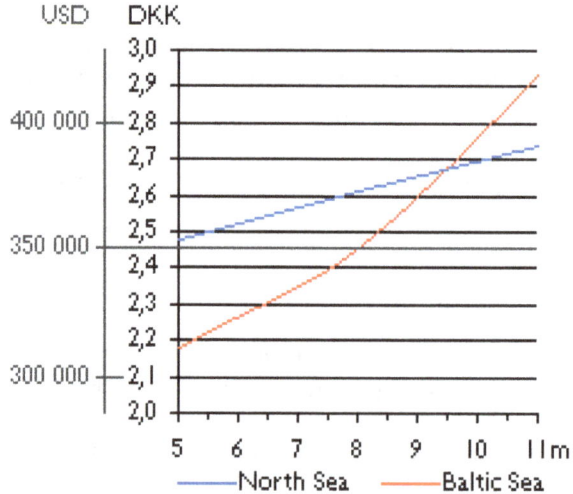

The dimensioning factor of the foundation varies from the North Sea to the Baltic Sea. In the North Sea it is the wave size that determines the dimension of the mono pile. In the Baltic Sea the pack ice pressure decides the size of the foundation. This is the reason why the mono pile foundation cost increases more rapidly in the Baltic Sea than in the North Sea. The costs include installation.

Erosion will normally not be a problem with this type of foundation.

References

- Underpinning-methods-procedure-applications-14480: theconstructor.org, Retrieved 28 May 2018

- 3-most-common-home-foundations-the-pros-cons: reinbrechthomes.com, Retrieved 15 March 2018

- Foundation-engineering, geotechnical-engineering, deep-foundation-35: civiltoday.com, Retrieved 16 July 2018

- Deep-foundation-and-pile-driving-construction-explained: coastalbridge.com, Retrieved 10 April 2018

- Rubble-trench-foundation, permahome: naturalhomes.org, Retrieved 25 April 2018

Soil Mechanics

The study of the behavior of soils, or analysis of the flow of fluids in natural and man-made structures as well as investigation of the deformation of structures that are supported by or made of soil are within the scope of soil mechanics. The aim of this chapter is to provide an insight into the field of soil mechanics, through topics such as frost heaving, soil shrinking swelling, bearing capacity, stresses in the ground, lateral earth pressures, soil porosity, soil degradation, etc.

The term "soil" can have different meanings, depending upon the field in which it is considered.

To a geologist, it is the material in the relative thin zone of the Earth's surface within which roots occur, and which are formed as the products of past surface processes. The rest of the crust is grouped under the term "rock".

To a pedologist, it is the substance existing on the surface, which supports plant life.

To an engineer, it is a material that can be:

- Built on: foundations of buildings, bridges
- Built in: basements, culverts, tunnels
- Built with: embankments, roads, dams
- Supported: retaining walls

Soil Mechanics is a discipline of Civil Engineering involving the study of soil, its behavior and application as an engineering material.

Soil Mechanics is the application of laws of mechanics and hydraulics to engineering problems dealing with sediments and other unconsolidated accumulations of solid particles, which are produced by the mechanical and chemical disintegration of rocks, regardless of whether or not they contain an admixture of organic constituents.

Soil consists of a multiphase aggregation of solid particles, water, and air. This fundamental composition gives rise to unique engineering properties, and the description of its mechanical behavior requires some of the most classic principles of engineering mechanics.

Engineers are concerned with soil's mechanical properties: permeability, stiffness, and strength. These depend primarily on the nature of the soil grains, the current stress, the water content and unit weight.

Formation of Soils

In the Earth's surface, rocks extend upto as much as 20 km depth. The major rock types are categorized as igneous, sedimentary, and metamorphic.

- Igneous rocks: formed from crystalline bodies of cooled magma.

- Sedimentary rocks: formed from layers of cemented sediments.

- Metamorphic rocks: formed by the alteration of existing rocks due to heat from igneous intrusions or pressure due to crustal movement.

Soils are formed from materials that have resulted from the disintegration of rocks by various processes of physical and chemical weathering. The nature and structure of a given soil depends on the processes and conditions that formed it:

- Breakdown of parent rock: weathering, decomposition, erosion.

- Transportation to site of final deposition: gravity, flowing water, ice, wind.

- Environment of final deposition: flood plain, river terrace, glacial moraine, lacustrine or marine.

- Subsequent conditions of loading and drainage: little or no surcharge, heavy surcharge due to ice or overlying deposits, change from saline to freshwater, leaching, contamination.

All soils originate, directly or indirectly, from different rock types.

Physical weathering reduces the size of the parent rock material, without any change in the original composition of the parent rock. Physical or mechanical processes taking place on the earth's surface include the actions of water, frost, temperature changes, wind and ice. They cause disintegration and the products are mainly coarse soils.

The main processes involved are exfoliation, unloading, erosion, freezing, and thawing. The principal cause is climatic change. In exfoliation, the outer shell separates from the main rock. Heavy rain and wind cause erosion of the rock surface. Adverse temperarture changes produce fragments due to different thermal coefficients of rock minerals. The effect is more for freeze-thaw cycles.

Chemical weathering not only breaks up the material into smaller particles but alters the nature of the original parent rock itself. The main processes responsible are hydration, oxidation, and carbonation. New compounds are formed due to the chemical alterations.

Rain water that comes in contact with the rock surface reacts to form hydrated oxides, carbonates and sulphates. If there is a volume increase, the disintegration continues. Due to leaching, water-soluble materials are washed away and rocks lose their cementing properties.

Chemical weathering occurs in wet and warm conditions and consists of degradation by decomposition and/or alteration. The results of chemical weathering are generally fine soils with altered mineral grains.

The effects of weathering and transportation mainly determine the basic nature of the soil (size, shape, composition and distribution of the particles).

The environment into which deposition takes place, and the subsequent geological events that take place there, determine the state of the soil (density, moisture content) and the structure or fabric of the soil (bedding, stratification, occurrence of joints or fissures).

Transportation agencies can be combinations of gravity, flowing water or air, and moving ice. In water or air, the grains become sub-rounded or rounded, and the grain sizes get sorted so as to form poorly-graded deposits. In moving ice, grinding and crushing occur, size distribution becomes wider forming well-graded deposits.

In running water, soil can be transported in the form of suspended particles, or by rolling and sliding along the bottom. Coarser particles settle when a decrease in velocity occurs, whereas finer particles are deposited further downstream. In still water, horizontal layers of successive sediments are formed, which may change with time, even seasonally or daily.

Wind can erode, transport and deposit fine-grained soils. Wind-blown soil is generally uniformly-graded.

A glacier moves slowly but scours the bedrock surface over which it passes.

Gravity transports materials along slopes without causing much alteration.

Soil Type

Soils as they are found in different regions can be classified into two broad categories:

- Residual soils
- Transported soils

Residual Soils

Residual soils are found at the same location where they have been formed. Generally, the depth of residual soils varies from 5 to 20 m.

Chemical weathering rate is greater in warm, humid regions than in cold, dry regions causing a faster breakdown of rocks. Accumulation of residual soils takes place as the rate of rock decomposition exceeds the rate of erosion or transportation of the weathered material. In humid regions, the presence of surface vegetation reduces the possibility of soil transportation.

As leaching action due to percolating surface water decreases with depth, there is a corresponding decrease in the degree of chemical weathering from the ground surface downwards. This results in a gradual reduction of residual soil formation with depth, until unaltered rock is found.

Residual soils comprise of a wide range of particle sizes, shapes and composition.

Transported Soils

Weathered rock materials can be moved from their original site to new locations by one or more of the transportation agencies to form transported soils. Transported soils are classified based on the mode of transportation and the final deposition environment.

Soils that are carried and deposited by rivers are called *alluvial deposits*.

Soils that are deposited by flowing water or surface runoff while entering a lake are called lacustrine deposits. Alternate layers are formed in different seasons depending on flow rate.

If the deposits are made by rivers in sea water, they are called *marine deposits*. Marine deposits contain both particulate material brought from the shore as well as organic remnants of marine life forms.

Melting of a glacier causes the deposition of all the materials scoured by it leading to formation of glacial deposits.

Soil particles carried by wind and subsequently deposited are known as aeolian deposits.

Frost Heaving

There was once a widely held belief among the rural inhabitants of several English counties that stones grew and multiplied. To the farmers and laborers who worked the land this explained why, when a field was cleared of stones, it would soon be littered with more. The common thought was that the stones grew from small pebbles in the soil and then somehow rose to the surface. Many people even went so far as to believe that the pebbles themselves were the offspring of so-called "mother-" or "breeding-stones," in reality chunks of an unusual siliceous flint conglomerate that is also known locally as "puddingstone" because of its resemblance to an old-fashioned plum pudding.

Puddingstone is a type of conglomerate in which the color of the clasts contrasts sharply with that of the matrix giving it the appearance of an old-fashioned plum pudding. The photo shows an example of Hertfordshire puddingstone (named after the county of Hertfordshire in southern England where most of it is found), which consists of well-rounded brown and dark gray flint pebbles distributed throughout a strongly cemented, much lighter-colored siliceous matrix. In some parts of England rocks like this were known as "mother-stones" or "breeding stones" in the belief that pebbles were their offspring.

Mother-stones are still kept by some, but only as curiosities and conversation pieces. No-one in their right mind today would seriously bestow them with any procreative capability or believe that pebbles are baby rocks. And the arrival of yet another multitude of stones that in years past country people could only attribute to some sort of magic, we now recognize as nothing more than one manifestation of a natural process called frost heave. It is the same process, often aggravated by flooding and/or heavy traffic, that ruins miles of roads every spring: the obstacle course of broken pavement, potholes, and carbuncle-like humps that they become as winter loses its grip and the annual thaw sets in is primarily the result of frost heave.

Frost heave is a form of frost action, a physical weathering process involving the cyclic freezing and thawing of water in soil or rock. Heave in this context refers to the upward movement of the ground surface that occurs in response to the seasonal formation of ice in the underlying soil. The dynamics of this ostensibly simple process are exceedingly complex and still not fully understood; yet despite its intricacies, the incidence of frost heave is dependent on just three things: besides freezing temperatures, it only requires the right kind of soil and an abundant, readily available water supply.

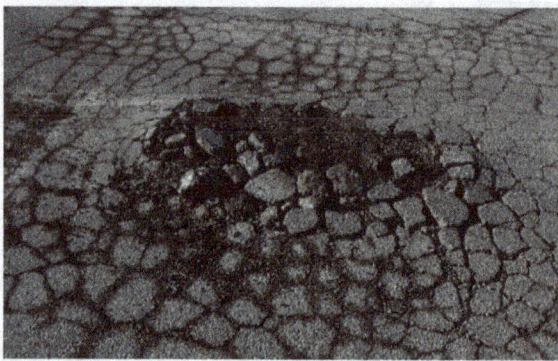

Asphalt pavement on a Bismarck street damaged by frost heave. Several chunks of the surface course have been dislodged by the wheels of local traffic exposing the underlying aggregate base.

Early investigators attributed frost heave to the expansion that occurs when water freezes, but this did not explain why the effects of frost heave are so frequently much greater than frozen soil alone is capable of producing. The discrepancy quickly becomes apparent if you consider that when water freezes its volume increases by about 9%. That being the case, a soil with a water content of 50% (about as much as the average soil can hold) would only expand by half this amount. Yet even if this expansion was unidirectional (the old theory assumed it was upwards because that was the path of least resistance) it would only produce about half an inch of uplift per vertical foot of frozen soil. In a typical winter the ground freezes to an average depth of about 4.5 feet (1.4 m) but as figure below shows, heaving can far exceed the 1 or 2 inches (2.5-5 cm) this premise forecasts. Clearly, something else is going on.

Frost heave on this Bismarck street has lifted parts of the road surface about 18 inches (46 cm). The maximum freeze depth in the Bismarck area for the winter of 2010-2011 was about 40 inches (100 cm).

The problem with the theory that frost heave is caused by the expansion of water when it freezes is that it was based on an erroneous assumption, which was that soils behaved as closed systems (nothing in or out). Evidence to the contrary remained largely ignored until Stephen Taber, a geologist at the University of South Carolina, published the results of a series of investigations that debunked the old theory once and for all. Taber demonstrated that it was not expansion, but rather the formation of ice lenses by segregation of water from the soil as the ground freezes that is the principal cause of frost heave. Moreover, by experimenting with soils as open systems he was able to show that lens growth may be sustained by the addition of groundwater drawn from warmer zones below the freezing front, and also that liquids other than water (Taber used benzene and nitrobenzene) can induce frost heave. This last observation clearly confirmed that the volumetric expansion of water as it turns to ice could not be the driving force behind the vertical displacement of frozen soil because, like almost all liquids besides water, benzene and nitrobenzene contract as they solidify.

Ice lenses are lens-shaped masses of almost pure ice that form in frozen soil or rock. Lens formation takes place at, or a short distance behind, the freezing front at any depth where conditions are favorable and will continue until those conditions change. Because they form perpendicular to the direction of heat flow, ice lenses tend to grow with their long axes oriented parallel to the ground surface. Most lensing is periodic, giving rise to multiple lenses separated vertically from one another by a layer of frozen soil. Single, thick lenses are rare in temperate climatic zones because their formation depends on persistent, steady state conditions. Ice lenses range in size from hair-thin slivers, barely visible to the naked eye, to laterally very extensive structures several feet thick.

An up-close examination of an ice lens reveals a structure composed of a multitude of thin, hair-like crystals that give the lens what has been described as a "satiny" appearance. The crystals develop, and are thus oriented; parallel to the direction of heat flow and it is their growth, not the path of least resistance that controls the direction of heave.

The ground freezes from the surface down, progressing along a front parallel to the surface and perpendicular to the direction of heat flow, thereby creating a thermal gradient as the ground loses heat to the cold air above it. An ice lens also forms from the top down and grows by the addition of water to its lower, warmer surface from soil below the freezing front. This is a far-from-straight-forward process because it requires a set of conditions that will allow water to (i) defy gravity and flow upwards and (ii) co-exist with ice at temperatures below freezing.

In a porous medium like soil, getting water to flow upwards is not difficult because it will rise naturally by capillary action. If the water is moving through unfrozen soil towards the freeze front, then this rise is aided by virtue of the fact that it is moving down the thermal gradient. Water in freezing soils is also moved around by cryogenic suction forces that are produced within a partially frozen region, or frozen fringe, just below the basal surface of the growing ice lens.

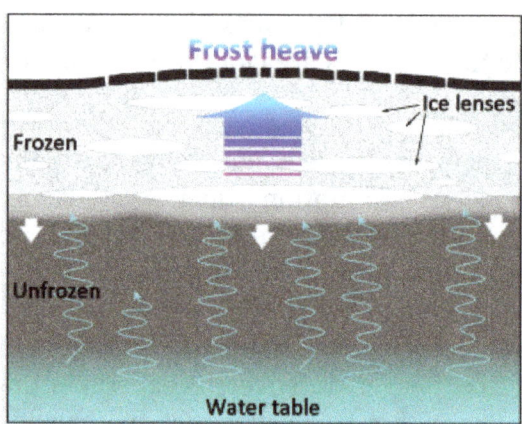

Formation of ice lenses in frozen ground overlain by pavement. Freezing takes place from the surface down (white arrows). When conditions are favorable, an ice lens begins to form a short distance behind the freezing front. Fed by liquid groundwater from warmer zones deeper in the soil profile, the lens will continue to grow for as long as the controlling mechanisms allow. Multiple lenses reflect fluctuations in these controls due to changes in surface temperature (which affects the rate of freezing), soil texture, and water availability. The amount of vertical displacement (heave) is roughly equal to the combined thicknesses of the underlying ice lenses. Heaving causes the pavement to crack, resulting in the kind of damage.

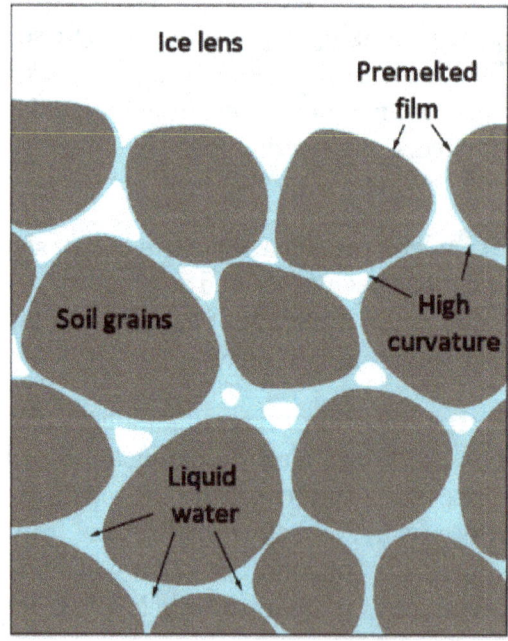

The frozen fringe below a growing ice lens (represented by the hazy area below the lowest ice lenses in figure above. Even though the temperature here is a few degrees below freezing, liquid water is able to exist in equilibrium with ice owing to (i) melting point depression caused by the Gibbs-Thomson effect and (ii) premelting at the ice-soil interface.

The microscopic interactions between soil grains, ice, and water that are responsible for these suction forces are also the fundamental controlling influence on ice lens formation and growth, and thus, ultimately drive the whole process of frost heave. They are also the reasons why water can sometimes remain in liquid form at sub-zero (0 °C) temperatures.

There are two thermodynamic processes that allow liquid water to persist at below-freezing temperatures. The first is the Gibbs- Thomson effect, which is a surface phenomenon that causes melting point depression. When ice freezes in a porous medium, solidification takes place along a curved interface that is convex-outward with respect to the ice. Freezing progresses by the penetration of this solid-liquid interface into increasingly smaller pore spaces; and in the same way that the radius of curvature of the meniscus formed by capillary action at a liquid-vapor boundary is proportional to the internal radius of the capillary tube, i.e., the narrower the pore space, the higher the convexity of the ice surface. At a particular sub-zero temperature a level of pore space confinement is reached where the energy required to further reduce the curvature of the ice-solid interface would exceed the amount required to maintain a small volume of water in its liquid form. As the temperature continues to fall, ice is able to move into more confined spaces and the amount of liquid water decreases until eventually it is all frozen.

Liquid water may also persist at below-freezing temperatures as a nanometers-thick, "premelted" film that forms along the surface contacts between individual soil grains and ice. Premelting is an effect that is surprisingly common in nature, and besides frost heave is known to have a controlling influence on a number of other natural phenomena including the electrification of thunderclouds, the formation of sea ice, and possibly even the development of planetary systems. The molecular interactions that cause premelting are complex, but the reason for the liquid films is simple enough: they are a barrier, set in place to attenuate the solid-solid electrostatic forces of repulsion between the surfaces of the ice and soil grains.

Water is transported along these liquid pathways from the zero isotherm to the base of the ice lens by pressure differences generated by the same forces that produce premelted films. The liquid film surrounding a soil grain within the frozen fringe is thicker on its lower (warmer) side. This variation in thickness is accompanied by a corresponding pressure change within the film, which establishes a pressure gradient across the soil grain parallel, but in the opposite direction, to the thermal gradient. The grain is thus pushed away from the ice lens towards warmer temperatures as water is drawn (sucked) through the film around the grain to feed the growing lens. The process is sustained by the continuous supply of liquid water from below the frozen fringe, and is an example of pressure-induced regelation.

Not all Soils are Susceptible to Frost heave

At about the same time that Taber was conducting his laboratory experiments in South Carolina, Gunnar Beskow, a geologist with the Swedish Institute of Roads, launched an investigation to address the costly problem of frost damage to Scandinavia's transportation network. In what was the first in-depth field study of frost heave, Beskow noted that the magnitude of its effects is largely determined by the particle size of the soil, an observation consistent with Taber's experimental findings.

They both found that the effects of frost heave are most pronounced in soils that facilitate capillary flow. Frost-susceptible soils thus tend to be fine-textured; with silts, loams, and very fine sands providing the optimum balance of moisture affinity (favored by high particle surface to volume ratios, i.e. small soil grains), pore size, permeability and hydraulic conductivity. Many glacial soils fall within this textural range, which is unfortunate because they quite literally cover a lot of ground Clays are hygroscopic, highly porous, and permeable but have low hydraulic conductivity because of their small void size. The consequent reduction in capillary flow tends to lessen the severity of heaving because it impedes lens formation. Frost heave is very rare in coarse-grained soils especially clean, well-drained sands and gravels. Their pore spaces are too large to permit capillary flow or the formation of ice lenses, so any pore water simply freezes in-place with no segregation.

Differential Frost heave, Freeze-thaw Cycles and Frost Jacking

Frost heave, if laterally uniform, is comparatively benign because there is no surface deformation and thus no threat to the integrity of pavements and other structures. Most frost damage is caused by non-uniform or differential heave. This occurs as a result of lateral variations in the conditions that control ice lens formation and other freezing behavior. The main influences are soil texture and the availability of water (i.e. the height of the water table). Other factors such as the presence or absence of vegetation, fluctuations in the thermal regime, and overburden pressure are also important but of lesser significance.

Load restrictions are imposed to prevent damage to roads weakened by
melt water saturation during the spring thaw.

Differential frost heave is what turns our roads into roller coasters every spring and causes the kind of damage shown in figures above. Heaving is more destructive at that time of year, not only because the freeze depth is at or near its maximum, but also because its effects are intensified by an influx of water from melting snow and ice, and the wear and tear the road has been subjected to all winter.

The ground thaws from the surface down, which can seriously impede soil drainage. Unless there is sufficient gradient to allow through flow, melt water will accumulate and remain trapped in the upper part of the soil until the layer of frozen ground beneath has completely thawed. A road with water trapped under it in this way can lose 50% or more of its normal load bearing capacity. This is why the Department of Transportation imposes load restrictions on certain roads for about a month after the onset of the spring thaw (figure), which is when they are at their weakest. Refreezing after a temporary thaw also exacerbates frost heave because the increased water content of the near-surface soil results in a correspondingly higher degree of segregation.

Shallow footings and foundations can crack under the influence of differential frost heave, which may render the structure they support unsound. Small buildings, especially if they are unheated, are particularly vulnerable to this problem, as are many other lightweight constructions such as decks, retaining walls, transmission towers, and so on.

The vertical displacement of an isolated structure (as opposed to part of a road or parking lot, for example) by frost heave is known as frost jacking, or frosts pull. But whereas a road lifted by frost heave will gradually subside to its original level after the ground beneath it has thawed, uplift caused by frost jacking tends to be permanent. The sequence of drawings in figure explains why.

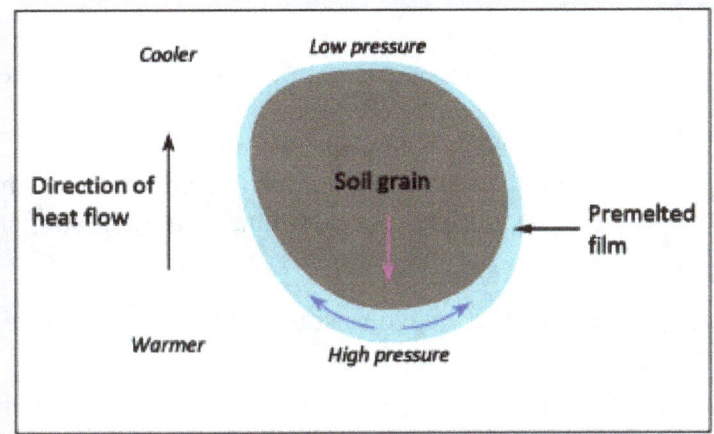

Segregation of soil and ice by regelation. The thickness of the premelted film surrounding a soil grain subjected to a temperature gradient is greater on its warmer side. Water in the premelted film migrates down the resulting pressure gradient (blue arrows) from the lower, warmer side of the grain to its upper, cooler side where the water refreezes, forcing the soil grain downwards (away) from the growing ice lens (red arrow).

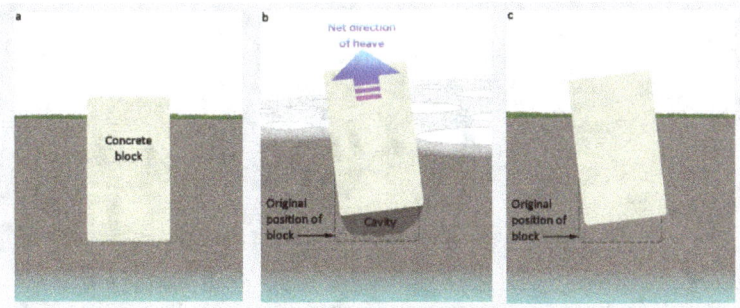

A cross-sectional view of a block of concrete partially buried in a moist, frost-susceptible soil. (a) Before the onset of freezing. (b) Frozen soil adheres to the sides of the block, forming bonds that become strong enough to overcome the downward shear and other resistant forces. Heave from growing ice lenses is transmitted to the block through these bonds, which lifts the block along with the surrounding soil, leaving a cavity that may collapse or fill with unfrozen debris. Frost jacking and differential heave commonly work in tandem, causing the block to tilt as it rises. (c) The block is unable to return to its original position when the ground thaws. If the cycle is allowed to repeat over several winters, the block will continue to be "jacked" upwards and possibly even forced completely out of the soil.

The effects of frost jacking are usually quite subtle but if the process is repeated over several winters, the cumulative damage can be significant. Properly observed, engineering guidelines and building codes designed to mitigate frost jacking (and other forms of frost heave) have proven highly successful; so reports of foundation piles being "jacked" 24 inches in a single year, or 11 feet over several winters should be a thing of the past. Nevertheless, there is no one-size-fits-all solution to frost heave, and what works well in one situation may fail miserably in another; so until we have a better understanding of the process, there are no guarantees.

Landforms Associated with Frost Heave

Frost heave is indiscriminate and anything we put in or on the ground, or which belongs there naturally (e.g. rocks and vegetation), is fair game. Railroads, bridges, pipelines, utility poles, coffins, unexploded ordnance, trash, small trees, fence posts, lawn edging, the list is endless and bizarre. As for the land itself, it is now plain to see that the annual appearance in spring of yet more stones on farmland cleared of them the previous year is due primarily to frost pull, probably aided a little by differential heave.

This seemingly uncanny ability of frost heave to sort unconsolidated material by size is taken to a whole new level in regions underlain by permafrost, where intense cold, recurrent diurnal and seasonal freeze-thaw cycles, and poorly drained soils all serve to enhance the processes of frost action. Differential heave is a key mechanism in the formation of some types of patterned ground.

Two examples of patterned ground. (a) Stone circles in the Svalbard Islands, northern Norway. (b) Frost boils (indicated by the arrows) in Abisko National Park, Lapland, Sweden.

The reorganization of surface materials into regular arrangements of various geometric shapes, commonly circles, polygons and lines (stripes). Sorted patterned ground, such as the stone circles in figure, develops on till and other diamictons. The stones are moved upward and outward by repeated freezing and thawing leaving the finer-grained material in the center of the ring. Frost boils (figure) are a type of non-sorted patterned ground consisting of circles of bare earth formed by a combination of frost heave and cryoturbation (soil disturbance caused by frost action). They are usually separated by areas of vegetation or peat and are not ringed by stones.

Large pingos near Tuktoyaktuk in the Mackenzie Delta in Canada's Northwest Territories. The one in the foreground is Ibyuk Pingo, Canada's highest at 161 feet (49 m) and the second-highest in the world. It is still growing at a rate of about three-quarters of an inch (2 cm) a year and is estimated to be at least 1,000 years old.

The small ice-cored hills known as pingos (from the Inuit word pinguq, meaning "little hill") are unique to the permafrost environment and their relict forms in modern temperate zones are indicative of an earlier, much colder climate. Viewed from above, pingos are either circular or elliptical in shape with basal diameters ranging from a few feet to more than a third of a mile. Up to 200 feet (70 m) high, they are generally rounded in profile although the summits of many of the larger ones are cratered and fractured, giving them the appearance of a small volcanic cone (figure). These ruptured surfaces are caused by tensile stress generated by the growing, typically massive ice core, which is often exposed as the displaced overburden of soil and vegetation slumps to lower elevations.

Most pingos may be classified as one of two types, based on their origin. Both types involve the formation of segregated ice, the principal difference between the two classes being the source of the water supply. Closed-system pingos develop in recently drained shallow lake basins, where liquid groundwater persists in a layer of unfrozen ground, or talik, within the former lake bed. Stripped of its insulating cover of lake water, the surface of the talik begins to freeze, completely enclosing the remaining water in permafrost. Hydrostatic pressure from the encroaching ice forces the water upward until it, too, eventually begins to freeze and heave the frozen overburden into a mound. Closed-system pingos are most common in areas of perennial permafrost and are generally the larger of the two type.

An aerial view of a group of well-developed palsas in northern Sweden.

Open-system pingos are mainly associated with discontinuous permafrost zones, where taliks and mobile ground water are common, typically forming on slopes where they obtain their water via artesian flow.

Palsas resemble small pingos, although unlike their larger counterparts, they are not limited to the permafrost environment and their genesis is different. Aptly named (palsa is a Finnish modification of the Northern Sami word balsa, meaning "a hummock rising out of a bog with a core of ice"), palsas typically occur in peat bogs and other wetlands as low, oval-shaped mounds that rarely exceed 300 feet (100 m) in length or rise much higher than about 30 feet (10 m). They consist of a core of alternating layers of segregated ice and peat or fine sediment that, insulated by a covering of vegetation, often remains perennially frozen, growing a little thicker each winter as more ice is added.

Whereas the core ice of pingos is derived from water under hydrostatic or hydraulic pressure, palsas are formed by cryosuction in more-or-less the same way as the hummocks that appear on our roads every spring. Besides the abundant water supply, their apparent affinity for boggy ground is due to bare spots where the wind has removed most or all of the snow cover, allowing frost to penetrate deeper into the subsurface than in the surrounding areas. Once a palsa has begun to form, its continued growth is favored by itsincreasing surface elevation, which the wind is more likely to keep clear of snow. Moreover, as the palsa rises higher above the water table, its covering of wetland vegetation starts to dry out, which makes it a more effectual insulator of the frozen core. An increase in surface albedo, brought about by a change in vegetation from peat mosses to lighter-colored plant species such as Cladonia (reindeer or cup lichens) may also influence palsa formation.

Heave is also the underlying cause of frost creep, the downslope movement of soil in response to cyclic expansion and contraction induced by freeze-thaw action. In combination with solifluction or gelifluction (its frozen-ground equivalent), frost creep operates on gradients as low as 1° but is most effective on 5°-20° slopes. Landforms are genetically similar and are usually classified morphologically. Features include lobes, benches, sheets, and streams.

Soil Shrinking Swelling

A soil survey contains maps and a description of each major soil in the survey area. More important, the survey describes how soil properties affect a wide range of rural and urban land uses. One of these properties, shrink/swell potential, is of great importance in the construction industry.

Shrink/swell potential is the relative change in volume to be expected with changes in moisture content, that is, the extent to which the soil shrinks as it dries out or swells when it gets wet. Extent of shrinking and swelling is influenced by the amount and kind of clay in the soil. Shrinking and swelling of soils causes much damage to building foundations, roads and other structures. A high shrink/swell potential indicates a hazard to maintenance of structures built in, on, or with material having this rating. Moderate and low ratings lessen the hazard accordingly.

Argentina as an Example

Soils have rigid behavior; that means relatively stable relations between their solid and pore volume. In the following figure of the three phase system, the solid phase remains stable while the volumes of water and air vary conversely within pore space.

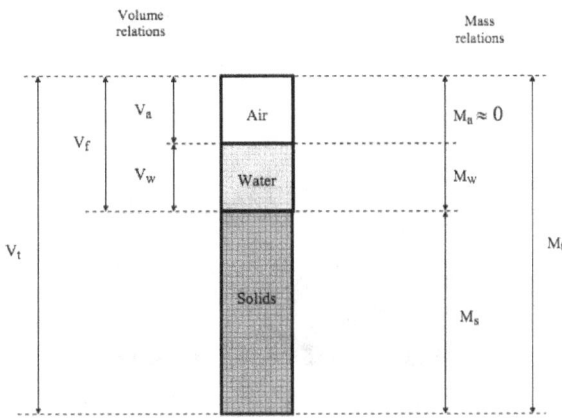

Schematic diagram of the soil as a three phase system

When a rigid soil dries, a given volume of air enters to the pore space in replacement of an equivalent volume of water. On the basis of this theoretical approach several methodologies have been developed. For instance, the determination of pore size distribution using the water desorption method.

$$\delta V_w = \delta V_a$$

Therefore, air-filled porosity increases as a rigid soil dries. As can be observed in figure below, rigid or non-swelling soils do not change their specific volume, v, and hence, their bulk density ρb during their water content θ variation range. Rigid or non-swelling soils are usually coarse – textured, organic matter – poor, and hard to till. They also have low aggregate stability, high module of rupture, and low resilience after a given damage (e.g. compaction by agricultural traffic). They are considered to have hard-set behavior. After several years of disc plowing, a hard plow pan is developed in the subsoil.

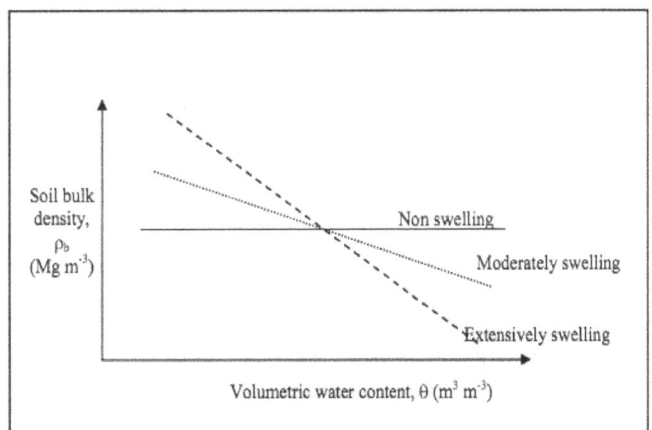

Schematic variation of soil bulk density in non-swelling (rigid), moderately swelling and extensively swelling soils.

In contrast, extensively swelling soils undergo significant bulk density, ρb, variations during their water content, θ, variation range. They are usually fine – textured, with smectitic type of clays. They develop desiccation cracks on drying, which confers them high resilience, and little tillage requirement. They are considered to have self-mulch behavior. Figure below shows an extensively swelling soil, a Vertisol cropped to soybean in the Argentine Mesopotamia. Their self-mulching facilitates their management under continuous zero tillage.

A hard plow pan developed in the subsoil of a sandy loam (Haplic Phaeozem) of the Argentine Pampas.

Extensively swelling Vertisol of the Argentine Mesopotamia, cropped to soybean using zero tillage.

Generally, most agricultural soils in the world develop only moderate volumetric changes during wetting and drying. This occurs provided the soil has less 8 % swelling clays. Although moderate, this swelling is highly important to the regeneration of soil structure after a given damage.

Processes Taking Place in Swelling Soils During Drying and Wetting

Different processes take place when a swelling soil dries or swells. On drying the soil decreases its volume by shrinkage, and desiccation cracks appear because of internal stresses in the shrunken and dried soil mass. These cracks are created in pre-existing planes of weakness within soil clods. As a result of shrinkage, soil decreases its height by subsidence. On wetting the soil increases its volume by swelling, the cracks are closed, and soil level rises. The process of swelling is mainly caused by the intercalation of water molecules entering to the inter-plane space of smectitic clay minerals.

The expansive characteristics of smectites are affected by the nature of adsorbed ions and molecules. Smectite increases its plane spacing as a result of the loss of adsorbed cations.

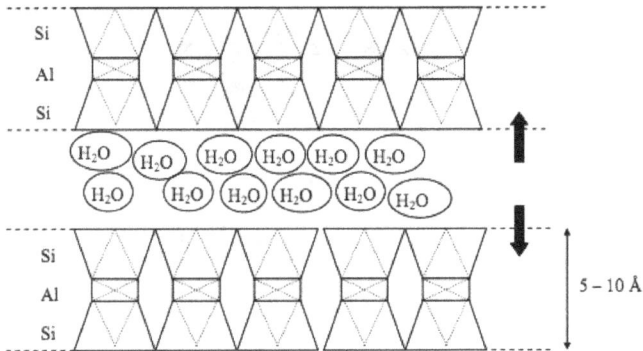

A diagram showing the intercalation of water molecules in the inter-plane space of clay smectites.

Types of Soil Swelling

When a dry soil wets, during the first stage it undergoes three dimensional (3-D) volumetric expansions, because its desiccation cracks are still opened. In a second stage, after desiccation cracks were closed, soil volumetric expansion is only 1-D, causing the rising of soil level,

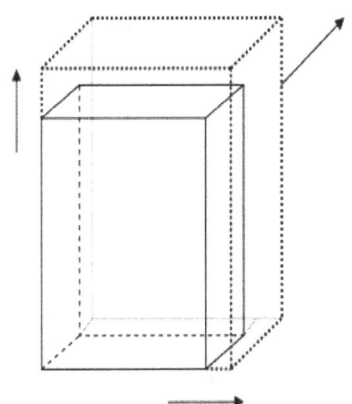

Schematic diagram of 3-D soil swelling

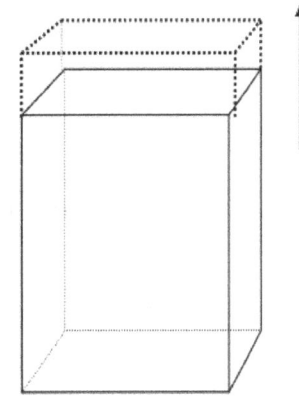

Schematic diagram of 1-D soil swelling

Consequences of Soil Swelling

Soil volumetric changes may cause both unfavorable and favorable effects on human activities. Unfavorable effects are the destruction of buildings, roads and pipelines in uncropped soils, and the leaching of fertilizers and chemicals below the root zone through desiccation cracks (by pass flow). In these soils horizontal cracks break capillary flux of water. On the other hand, swelling clays can be used to seal landfills storing hazardous wastes. This sealing avoids the downward migration of contaminants to groundwater. In cropped soils, the development of a dense pattern of cracks on drying improves water drainage and soil aeration, and decreases surface runoff in sloped areas. Soil cracking is closely related to the recovery of porosity damages by compaction. Here in, a conceptual model describing the sequence of paths leading to the development of desiccation cracks is provided. Tensile stresses are developed on drying, which to the creation of primary, secondary and tertiary cracks. The cracks are, at the same time, future void spaces, and represent the walls of the future aggregates. This sequence of paths is believed to recover soil porosity in a previously compacted soil layer.

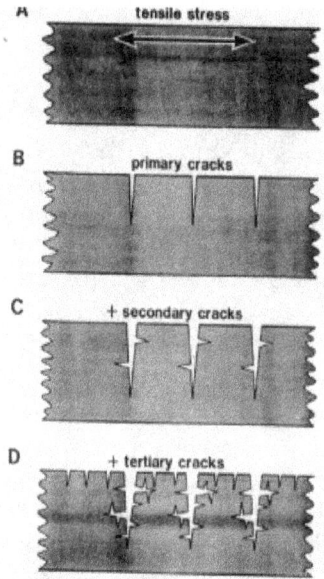

Conceptual model describing the development of primary, secondary and tertiary cracks, resulting from the buildup of tensile stresses on drying.

Methods for Assessing Soil Swell-shrink Potential

a) Coefficient of linear extensibility, COLE

It characterizes the variation of soil volume from 1/3 atm water retention (i.e. field capacity) to oven dry conditions:

$$COLE = \left(v_{1/3atm} - v_{dry} \right)^{1/3} - 1$$

where $v_{1/3atm}$ is the soil volume at $1/3\,atm$ water retention and v_{dry} the soil volume at oven dry conditions. According to their COLE, a range of soil swell-shrink potential can be distinguished:

Soil swell – shrink potential : COLE

Low	< 0.03
Moderate	0.03 – 0.06
High	0.06 – 0.09
Very high	> 0.09

The COLE index was found to be closely related to a number of soil variables. The higher determination coefficients correspond to the contents of total and swelling clays.

Independent variables	R^2
Clay $< 2\,\mu$m	0.87 ***
Clay $2 - 0.2\ \mu$m	0.41 **

Clay < 0.2 μm	0.43 **
Smectites < 0.2 μm	0.61 **
interstratified swelling clay < 0.2 μm	0.2
swelling clay < 0.2 μm	0.91 ***
Porosity	0.33 *
Organic C	0.08
Sodium Adsorption Ratio	0.01

The Shrinkage Characteristic or Shrinkage Curve

Each soil has a characteristic water retention curve, which relates its water content to the energy at which water is retained by the solid phase (soil matric potential). Likewise, swelling soils may be also characterized by its shrinkage curve. This shows the variation of soil specific volume, v, with water content, θ, during the air – drying of water saturated natural clods.

To construct a shrinkage curve, the volume of natural soil clods must be determined by hydrostatic up thrust in a non-polar liquid (kerosene) during air drying. The corresponding water content is also measured at different times of drying. Another option is to coat the clods with SARAN resin, and then, submerge them in water.

In a shrinkage curve the inverse of bulk density (i.e. soil specific volume, v) is plotted to the volumetric water content θ of the soil. In this graphic two theoretical lines are depicted: a) the solid phase line (from the converse of soil particle density) that represents the lowest soil volume of a soil having zero pore space; and b) the 1:1 saturation line that represents soil swelling with zero air within pore space.

Figure: Experimental device to measure the volume of natural clods,
by hydrostatic up thrust in a non polar liquid (kerosene).

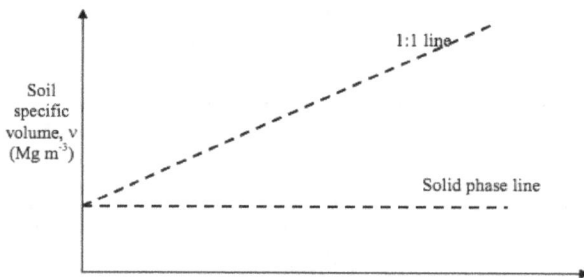

Figure: Initial plot to construct a soil shrinkage curve

After introducing v – θ pairs of data, straight lines can be fitted. This allows the identification of different shrinkage zones. Normal shrinkage (B → A) is characterized by equivalent decreases in both v and θ on drying, and thus, no air enters into soil pores. In the drier range of the θ variation, soil v decreases during drying are lower or even null. Residual shrinkage (A → α) allows air entrance into soil pores, and hence the creation of air-filled porosity. Swelling soils are considered to have normal, or equivalent v and θ variations throughout their water variation range. Moderately swelling soils, in turn, develop residual shrinkage during the drier range of soil moisture variation. This allows air entry to soil pore space, and the process is also recognized as irreversible shrinkage. The location of the air entry point θ_A is considered a index of soil quality in swelling soils, the higher θ_A, the better the soil aeration.

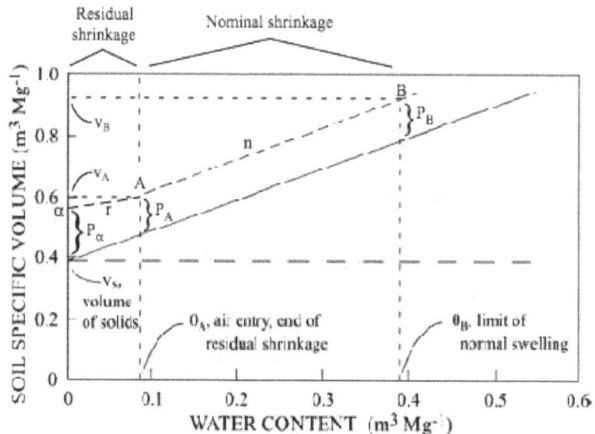

Figure: Theoretical shrinkage curve of a swelling soil.

Table: indices and related variables from the shrink data of natural soil clods.

θ_B	θ at the limit of normal swelling
θ_A	θ at the air entry point, i.e. the end of residual shrinkage
n	slope of the line B → A (normal shrinkage)
r	slope of the line A → α (residual shrinkage)
v_B	specific volume at the limit of normal swelling

V_A	specific volume at the air entry point
α	specific volume at zero water content
P_B	specific volume of air filled pores at B
P_A	specific volume of air filled pores at A
P_α	specific volume of air filled pores at α
$\theta_B - \theta_A$	difference between θ at the limit of normal swelling
	and θ at the air entry point, i.e. range of θ in the normal shrinkage zone

Mc Garry and Daniells derived several indices and related variables from the shrink data of natural soil clods. These are derived from the mathematical expression of two or three straight lines, fitted to the data. The figure shows the different soil shrinkage lines and indices of interest. We applied this approach to study soil volumetric variations in the Pampean soils of Argentina.

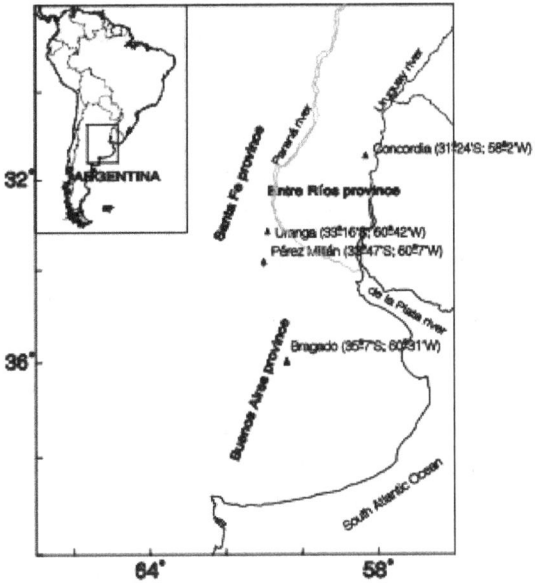

Figure: Geographical location of the main Argentine cropland area. Vertisols are highly conspicuous in Entre Ríos province, while silty loams covered thousand hectares in Santa Fe and Buenos Aires provinces.

Shrinkage characteristic of pampean silty clay loams affected by water erosion:

Silty loams (Argillic Phaeozems) are highly conspicuous in the north of the Pampean region. These soils are affected –to a different degree- by degradation, which can be classified in moderate and severe. Figure below graphically describes how degradation changed soil properties in the moderate and severe levels. Because of their high content of fine silt (2-20 µm), these soils have low structural regeneration capacity after degradation. We hypothesized whether this could be improved, or not, by the enrichment of topsoil with swelling clays. This situation can be found in severely degraded soils, in which the shallow A horizon was previously mixed with the below lying B horizon.

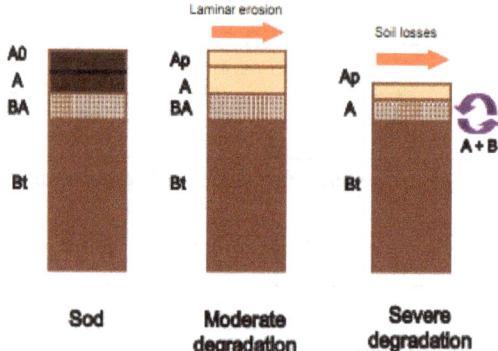

Figure: Idealization of soil profiles in non-degraded (sod), moderately and severely degraded situations.

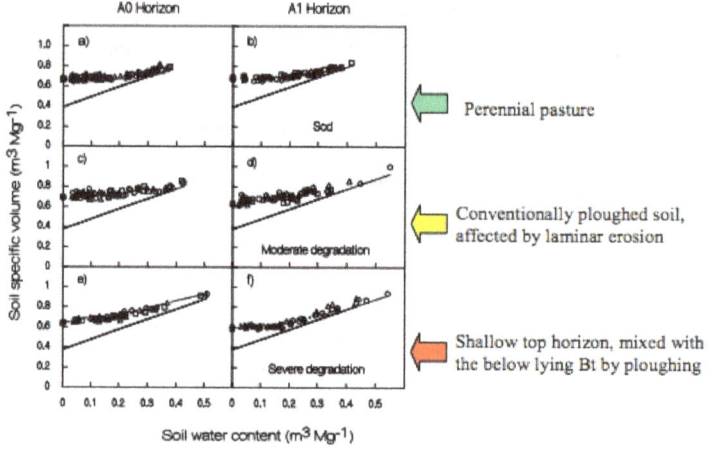

Figure: Soil shrinkage curves in the A0 and A1 horizons of a Peyrano silty clay loam
(Argillic Phaeozem), under different degradation levels.

Soil shrinkage curves differed because of the different degradation levels. At first sight it becomes evident the wider volumetric variation range of the severely degraded soils. Soil volume at zero water content, α, and the air entry point θ_A showed no significant differences between the sod and moderately degraded soils, and were significantly lower in the severely degraded soil. The slope, n, increased significantly from the sod to the severely degraded soil. The same happened with the normalcy range. The severely degraded soil reached significantly higher volume, v_B, and water content θ_B when swollen at maximum.

It can be concluded that clay enrichment (severe degradation) accentuated soil swelling (> v_B and θ_B and normalcy range), but did not improve air filled porosity. Soil horizons mixture cannot be recommended to the farmers, as a practice suitable to improve topsoil structure in Pampean silty loams.

Shrinkage characteristic of natric soils in the Flooding Pampa:

In the flooding Pampa there are different kinds of Solonetzes, that are periodically flooded. The region is characterized by water table rises and surface ponding during winter-spring periods. Figure below (a through d) shows soil specific volume – water content relations in two different

Solonetzes of the region. It can be observed that the fitted straight lines departed from the 1:1 line, showing air – filled porosity increases as soil wets.

The soils have no definite expansible clay mineralogy. Water table rises and surface ponding promote the buildup of trapped air pressures in top horizons. This shows abnormal soil swelling caused by air entrapment.

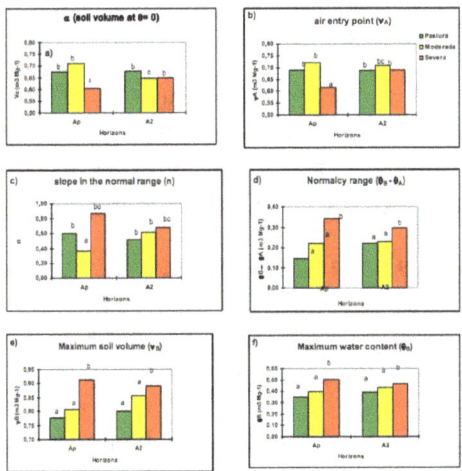

a) Soil volume at zero water content, α; and b) air entry point θ_A in the sod, and moderately and severely degraded soils. c) Slope in the normal range, n, and d) the normalcy range, $\theta_B - \theta_A$, in the sod, and moderately and severely degraded soils. e) Maximum soil volume, v_B, and d) maximum water content, θ_B, in the sod, and moderately and severely degraded soils.

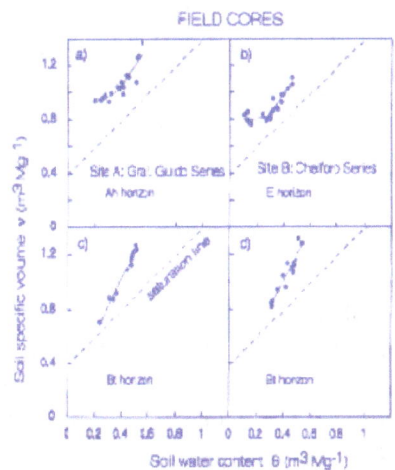

Figure: Soil specific volume, v, - water content relationships from repeated core sampling of surface (a, b) and Bt (c, d) horizons of two Solonetzes.

In the environmental conditions of the flooding Pampa of Argentina, trapped air is responsible for most of the swelling of soils having little expansible mineralogy of clays. Results show that trapped may be an important soil swelling factor in soils.

Shrink Swell Test

Expansive soils have long been recognized as important problem soils in geotechnical engineering. The significance of these soils throughout the world became prominent in 1965 in a conference that was to become the first in a series of seven conferences between 1965 and 1992. In 1995, the focus of these conferences shifted slightly toward the broader topic of unsaturated soils, and although the attention given to truly expansive soils may seem to have diminished, active research is still recorded in many parts of the world.

Despite 40 years of active research, expansive soil engineering remains one of the most inconsistently treated areas in international geotechnics. There are many different countries around the world where expansive soils are an issue, and almost as many different approaches used to treat them. Indeed, even within large countries like the USA, expansive soil engineering practice varies on a State by State basis.

In most cases, the testing methods used to assess the expansiveness of clay soil have been developed to obtain parameters for expansive soil behavioural models. The approaches that have been developed to estimate soil volume changes can generally be described as being based on either one-dimensional oedometer tests, or on soil suction/moisture content tests.

A third group, which infers expansive potential on the basis of physiochemical properties, such as molecular adsorption or cation exchange capacity have received much less attention.

Many of the approaches employing results from oedometer type tests have been developed in particular geographic regions with specific expansive soil problems, although they appear to be applicable in general situations. Most have focused on the prediction of heave using a constitutive theory based on changes in effective stress, soil suction, void ratio, and oedometer testing methods. Early models include Bishop, Croney, Coleman, Richards, Bishop and Blight, Burland , Matyas and Radhakrishna, Barden, Madedor and Sides, Aitchison and Woodburn, and Aitchison and Martin. These models were variously developed in terms of one or two different stress components, including mean total normal stress, matric suction, nett normal stress, and the deviatoric stress. Application of the more sophisticated models required the estimation of up to six compressibility parameters. Fredlund and Morgenstern considered the use of nett normal stress and matric suction as state parameters for an unsaturated soil, but from the perspective of multiphase continuum mechanics. They invoked a constitutive surface describing the void ratio of a soil in terms of the nett normal stress and the matric suction. The parameters required to define this constitutive surface are derived from a variety of tests including simple consolidation tests on saturated samples, suction measurements, and suction control using a pressure plate apparatus, shrinkage-type tests, free-swell, and constant volume oedometer tests. Because of hysteresis, rigorous application of the approach required two sets of coefficients to be determined for situations involving wetting and drying, or loading and unloading of soil.

Alonso, Gens, and Hight proposed an alternative frame-work to describe volume changes in non-swelling unsaturated soils. Gens and Alonso extended this general framework to include expansive soils. However, application of this model requires relatively complicated laboratory testing that involves simultaneous control of both matric suction and normal stress.

A variety of oedometer type tests were developed to determine parameters for the various volume change models listed above. Among the earliest of these was the double oedometer method of

Jennings and Knight. It employed the results of "free swell "and "natural water content" oedometer tests, carried out on sample pairs, with one sample consolidated from field moisture content, and the other, consolidated after first being swelled to saturation under a token load. This test gives an estimate of a parameter that can be used in the above constitutive frameworks that is believed to be free of the effects of sample disturbance. Predictions of heave using the double oedometer test are generally thought to be satisfactory. The double oedometer test was simplified by Jennings to testing of a single sample. Sullivan and McClelland proposed an approach based on a "constant volume" test. A laterally constrained sample is in undated after being preloaded to overburden stress, with subsequent volume changes being restrained by applying further load, and then the sample is rebounded after the maximum pressure is reached. The maximum applied pressure is defined as the swelling pressure for the soil, for the initial moisture content and density of the soil.

A large variety of other oedometer-based heave prediction approaches are also reported in the literature. These are summarized by Nelson and Miller and will not be discussed further here. They include the Noble method, the Navy method, and the USBR method. The main criticism of most of these methods is also a failure to account for the effects of sample disturbance.

Volume change predictions based on suction (or moisture content) tests are an alternative to oedometer test based approaches. Ina comparison with oedometer test methods, Johnson found that suction test methods were simpler, more economical, and more efficient. Suction test based approaches typically use swelling or shrinking tests on undisturbed samples to relate changes in suction to changes in volume. As such, they are typically less rigorous and are most suited to situations where the nett stress does not change, such as in the prediction of ground movements due to moisture changes, under constant, mean total normal stress conditions.

One of the simplest approaches arises from a consideration of the basic relationship between volumetric strain, ε_v, water content, ω, initial void ratio, e_o, particle density, ρ_s, and water density, ρ_ω, in a saturated soil, as defined by the equation:

$$\varepsilon_v = \frac{\rho_s}{\rho_\omega} \frac{\Delta\omega}{1+e_o}$$

where $\Delta\omega$ is the change in water content from the initial void ratio condition.

An equation similar to above equation was employed in heave predictions by Richards, who included a factor to account for the effects of lateral restraint. The term $\dfrac{\rho_s}{\rho_\omega(1+e_o)}$ effectively represents a parameter relating changes in moisture content to changes in soil volume. Hanafy also derived equations similar to above equation to relate changes in water content to changes in void ratio. Hanafy noted from his experiments that the proportionality of volume strains with moisture changes was only applicable (approximately) to the moisture range between the shrinkage limit and the equilibrium moisture content, where the equilibrium moisture content was defined as the moisture content beyond which no significant void ratio changes take place. Two further limitations of the approach embodied in equation above are that the moisture content—volume change relationship breaks down in dry soil conditions, and secondly, there is no consideration of the influence of loading on volume change.

The WES method of the U.S. Army Corps of Engineers relates volume changes to changes in suction. This method employs an analysis in the plane of constant net stress, using soil suctions measured by transistor psychrometer, as well as suction indices, calculated from estimates of the slope of the specific volume water content curve of a soil. A similar approach was also adopted by Dhowian.

Another commonly used suction based test method is the CLOD test. This method employs an un-restrained shrinkage test on an undisturbed soil sample, using a resin coating technique. Irregular lumps of soil can be used. The volume of the sample is determined as a function of its moisture content as it dries, providing a volume change index. The method has been found capable of providing good ground movement pre-dictions, but is limited to situations where ground movements occur under relatively constant nett stress conditions.

Although the above suction/moisture based methods could reasonably be adapted to estimate volume changes in both shrinking and swelling soils, the emphasis in their development and use appears to have been mainly placed on the prediction of heave due to wetting soil profiles.

The behavior of expansive soils subjected to successive or cyclic shrink and swell events has been looked at by a number of researchers. However, such studies have primarily been undertaken for the purposes of examining fundamental soil behavior under such conditions, and not in the context of establishing a method for the routine evaluation of expansive soil potential.

A number of approaches have specifically been developed, more generally, to enable prediction of both swelling (and heave) and shrinkage in clay soil profiles. Hanafy, realizing that measurements of expansive potential from shrinkage or swelling tests may be significantly biased by the natural water content of the sample, proposed a test involving both free swell and free shrink components. He suggested that the almost linear relationship between water content changes and void ratio changes over a reasonable range of moisture content could be used in the prediction of ground movements, although the effectiveness of the approach was not demonstrated by application to real soils. Alternatively, methods employing suction are based on the observation that changes in log suction are approximately linearly proportional to changes in soil volume, over the range of suction change commonly experienced under field conditions. The heave prediction method of McKeen, based on the CLOD test, recognized this linearity and exploited it by defining its slope as an index (the "total suction-water content index") that could be used to predict soil volume changes.

In other approaches, the constant of proportionality was referred to as an instability index, I_{pt}. In current Australian practice, the term instability index has been given a slightly different definition, although it depends upon the measured soil vertical strain for a unit change in total soil suction for a laterally unconfined sample, a property termed the shrinkage index, I_{ps}. I_{pt} is derived from I_{ps} by taking account of specific factors of the physical field situation, such as lateral stress and confinement effects. Both indices are termed "reactivity indices," as the vertical movement may be either shrinkage with increase of suction, or swelling with loss of suction.

There are a variety of simple approaches based in the instability index approach. Aitchison and Martin followed the approach of Richards, and showed that in assuming that the change in water content is approximately linearly related to the change in the log of the soil suction, the approach of equation above given is consistent with the instability index approach. If "c" is an experimentally determined constant of proportionality between gravimetric moisture content, ω, and the logarithm of suction and ψ is a suction variable, then "c" can be expressed as,

$$c = \frac{\Delta\omega}{\Delta\log(\psi)}$$

If the relationship between the swelling index, I_{ps}, the vertical strain component, ε_z, and the corresponding change in log suction is defined by,

$$\varepsilon_z = \frac{\varepsilon_v}{3} = I_{ps}\Delta\log(\psi)$$

then eliminating the water content from equation $\varepsilon_v = \frac{\rho_s}{\rho_\omega}\frac{\Delta\omega}{1+e_o}$ gives,

$$I_{ps} = \frac{1}{3}\frac{\rho_s}{\rho_\omega}\frac{c}{1+e_o}$$

The factor of one third in equation $\varepsilon_z = \frac{\varepsilon_v}{3} = I_{ps}\Delta\log(\psi)$ is a nominal correction factor, included to give the index in terms of the vertical strain component, assuming that the vertical strain in an unconstrained sample is one third of the volumetric strain.

Cameron reviewed a number of different tests for the measurement of swelling indices, I_{ps}, for the direct estimation of shrinkage indices for implementation in equations of the type of equation $\varepsilon_z = \frac{\varepsilon_v}{3} = I_{ps}\Delta\log(\psi)$. One of these is the core shrinkage test as described by Mitchell and Avalle. In this test, a small, undisturbed core of clay soil is trimmed to a diameter of 38–50 mm, and a length of twice its diameter. It is then air dried for two days before being oven dried. Regular measurement of the sample mass and length are made through out. Also, the suction of the sample at its initial moisture state is measured and a soil water characteristic curve is determined. Using the data corresponding to the initial linear part of the drying curve, a swelling index referred to as the core shrinkage index, I_{cs} is determined as follows,

$$I_{cs} = \frac{\varepsilon}{\Delta\log(\psi)} = \frac{\varepsilon}{\Delta\omega}\cdot\frac{\Delta\omega}{\Delta\log(\psi)} = \frac{\varepsilon.c}{\Delta\omega}$$

Another simple test is the loaded shrinkage test. In this test, a small, undisturbed sample is placed in a perforated shrinkage cell under a nominal load of 25 kPa (to simulate the pressure of a typical lightly loaded foundation). The initial moisture content and suction are determined. The sample is then subjected to controlled shrinkage by being allowed to reach mass equilibrium in the airspace above a supersaturated copper sulphate solution in a vacuum desiccator, which produces an equivalent soil suction of 4.5 pF units or 3.1 MPa (pF is defined as log negative [hydraulic head in centimeters], or by 1.01 + log[suctionin kPa]). After corrections are applied to the measured data for the initial load settlement of the sample, a reactivity index referred to as the loaded shrinkage index, I_{ls}, is determined using an equation similar to equation $I_{cs} = \frac{\varepsilon}{\Delta\log(\psi)} = \frac{\varepsilon}{\Delta\omega}\cdot\frac{\Delta\omega}{\Delta\log(\psi)} = \frac{\varepsilon.c}{\Delta\omega}$.

The controlled shrinkage test can take more than 8 weeks to achieve equilibrium of the small samples used.

The first quantitative assessments of soil expansiveness in Australia are thought to have been carried out in the late 1960s. Despite the growing awareness of expansive soils that arose in this era, the use of such assessments to guide engineering practice were extraordinary in the 1970s, and where performed, they were based on simple core shrinkage tests, and guided by work such as that of Aitchison and Woodburn and Aitchison and Martin. In the early 1980s, the need for improved engineering of expansive soil foundations was becoming apparent, and the assessment of soil expansiveness through specific testing was being undertaken on an infrequent basis by the consulting industry, mostly using core shrinkage methods such as that described in Mitchell and Avalle.

The use of a core shrinkage test was considered inadequate by many, due to its sensitivity to the initial moisture content of the sample. Samples that were unusually wet at the time of collection allowed measurement of shrinkage over most of their "working range" of moisture content, and so the values obtained were considered adequately representative. Samples that were unusually dry at the time of collection shrank little, if at all, and so the results were often unreliable. Cameron also found in a later study that the reliability of the core shrinkage test was strongly dependent on the moisture content-suction relationship for the soil, which could not be either simply determined or reliably derived by empirical relationships over a wide range of soil types.

Colin Thorne, a senior consulting engineer with Coffey Partners, is understood to have been the first to propose a shrink swell test. He proposed that a swell test be carried out in parallel with the shrink test, so that volumetric strains could be measured regardless of the initial moisture content. His idea was simply to add the axial strains measured from an unrestrained shrinkage specimen (ε_{sh}) and axial swell strains measured from a simple odometer specimen (ε_{sw}), both tested from the initial water content. Two reports by Colin Thorne, issued through Coffey Partners to the New South Wales Builders Licensing Board, are also considered to have helped shift the focus of foundation design in eastern Australia for lightly loaded structures away from settlement, and onto soil expansiveness. By this time, the need for better engineering in expansive soil foundations had become so obvious nationally that a committee was convened to prepare a draft Australian Standard to provide necessary guidance.

During the deliberations of this committee, the shrink swell test, as it currently exists, was formulated. It was considered that the axial swell strain from a 1D consolidation test, and the axial shrinkage strain component from an unrestrained core shrinkage test, were fundamentally different quantities, and that they could not simply be added together to produce a consistent result. It was decided that the swell component should be corrected to account for lateral restraint, by assuming that the suppressed lateral expansion would be redirected (to a greater or lesser extent) into the vertical direction. It was decided that a realistic swelling strain could be estimated (corresponding to a sample allowed to swell freely in all directions) by dividing the swelling strain measured in the restrained test by a factor of 2.0.

In order to derive a swelling index from this axial strain measurement, the suction change, corresponding to the variation in the measured axial strain, needs to be estimated. The use of parallel shrinkage and swelling tests meant that in a ll cases, the tests were conducted between the ranges of oven dry and effective saturation. It was considered, in the experience of the committee, that the significant, linearly varying portion of the strain-suction change relationship had a consistent suction range of 1.8 pF. It was thus decided that the test could be made attractive to routine geotechnical practice by adopting this value as being appropriate for all soils, in all tests, and hence, circumvent the need to determine suctions through direct measurement. The shrink swell index is thus defined as,

$$I_{ss} = \frac{\varepsilon_{sh} + \dfrac{\varepsilon_{sw}}{2}}{1.8}$$

The units of the shrink swell or reactivity index are %strain/pF of suction change.

In order to derive the instability index from the reactivity index, the index must be adjusted to account for the effects of surcharge and lateral confinement that act on the soil in its in situ condition. This is done by application of a factor, α, usually applied at the time at which ground movements are being estimated.

$$1_{pt} = \alpha . l_{ss}$$

The committee considered that in a cracked clay soil there is no lateral confinement, and so, the l_{ss} parameter should give a good estimate of the vertical soil strain component in an effectively unrestrained soil. Therefore, a value of α of unity was adopted in this zone. It was assumed that the effects of surcharge could be ignored at the shallow depths where shrinkage cracking occurs. In the soil below the cracked zone, lateral confinement exists, and it was considered that expansion in the vertical direction should be increased, but that below this depth, an increase in vertical swelling would be tempered by the surcharge restraint due to overlying soils.

In order to be consistent with the definition of l_{ss}, the effects of lateral restraint could be accounted for by using a value of $\alpha = 2$. This effectively reverses the correction applied to the swell component in determining the instability index, and applies a reversed correction to the shrink component. The effects of surcharge were more difficult to take account of. It was decided that the swell test could be carried out under a nominal applied load of 25 kPa, and that the effects of surcharge pressure could be included when the measured is values were used in a ground movement prediction. The methodology for this was developed by making recourse to anecdotal experience. Observations from the black soil plains of Moree in NSW, Australia, suggested that no movement occurred below 10 m, whilst at Vicksburg, Mississippi, it was similarly observed that "at 30 feet, clays don't swell." It was decided that the applied correction should result in no movement at a depth of 10 m, and that the effects of surcharge at any depth, z, should be linearly interpolated from total swell suppression at 10 m ($\alpha = 2.0$) depth, to no suppression at the surface ($\alpha = 0.0$). The resultant adjustment factor was thus defined as

$$\alpha \begin{cases} 1.0 & \text{creacked zone} \\ 2.0 - \dfrac{z}{5} & \text{uncracked zone} \end{cases}$$

Other authors such as McKeen have proposed theoretical α values. McKeen effectively incorporated an α factor as the product of two factors, f and s. Factor f was the lateral restraint factor, which is primarily related to the coefficient of earth pressure at rest, K_o, a s follows:

$$f = \frac{1 + 2k_o}{3}$$

Factor f was reported to lie within the range 0.50–0.83.

McKeen's second factor, s, was a load effect factor, which was assumed to depend upon the percentage of applied pressure to the swelling pressure, %SP. The relationship for %SP less than or equal to 50 % was:

$$s = 1.0 - 0.01(\%SP)$$

Mc Keen did not consider the influence of shrinkage cracking upon the ability of the soil to change volume. It should be noted also that factors f and s were applied to a reactivity index from CLOD tests, which was based on volume strain, and not vertical strain.

As described previously, the shrink swell test is composed of companion core shrinkage and swelling tests, carried out on undisturbed soil samples from their initial field moisture contents. The vertically oriented sample is usually obtained from the ground, using a 50-mm-diameter thin-walled tube. The sample is extruded from the tube as a soil core and a suitable portion of the sample is selected for the preparation of a shrinkage core (70–100 mm long) and a swell core (20–25 mm long). Test samples must be from adjacent portions of the core to ensure that water content and both compositional and structural differences are minimized. The shrinkage and swell tests are then conducted simultaneously, as follows.

Shrinkage Test—This component of the test is identical in procedure to the core shrinkage test, although fewer measurements are required as the shrink-swell index is based on the oven-dried state. A shrinkage core, 45–50 mm in diameter and a length of 1.5–2 diameters, is trimmed from the soil sample. Where possible, it is selected and trimmed to be free of major structural defects and loose material. Initial dimensions and mass are recorded. Small pins are added to each end as reference points to facilitate consistent measurements of sample length as drying proceeds. The shrinkage core is firstly air-dried. Regular measurements of length and mass are taken until shrinkage ceases. The core is then oven-dried to a constant mass at 105–110 °C, and final mass and dimensions are recorded. The data recorded enables initial and final water contents, axial strain to be calculated, and a graph of axial strain against water content to be plotted. Throughout the drying process, the core is kept in a shallow tray, so that any crumbs that become loose during the test are not lost, as this would affect moisture content calculations.

Swelling Test—This involves a simplified oedometer test in which the sample (of measured mass) is installed in a rigid ring, (of measured volume; usually around 20 mm high and 40–45 mm in diameter) and placed between porous stones in a consolidation apparatus. A gage to monitor the sample height is then zeroed under a nominal seating pressure of 5 kPa. A load of 25 kPa or the estimated in situ over burden pressure (whichever is greater) is then applied for 30 min to record any initial settlement or seating adjustment. This displacement is used to correct the initial sample height for determination of swelling strain. After re-zeroing the displacement gage, the sample is inundated with distilled water and allowed to swell until the swelling increment, in a period of not less than 3 h duration, is not more than 5 % of the total recorded swell. The initial water content is determined from the sample trimmings, and the final water content is measured from the extracted sample at the end of the test. In the sample preparation process, particular care is taken to ensure that the sample neatly fills the sample ring, as voids and recompacted or remolded portions will accommodate internal adjustments in the volume of the sample and hence, affect the realized vertical swell.

Shrinkage strains (ε_{sh}) and swell strains (ε_{sw}), measured in the respective tests, are then combined to give a swelling (shrink-swell) index, I_{ss}, according to equation $I_{ss} = \dfrac{\varepsilon_{sh} + \dfrac{\varepsilon_{sw}}{2}}{1.8}$.

Application of the Shrink Swell Test

The shrink swell index gives a quantitative measure of the vertical strain that will occur in a clay soil per unit change in suction. Thus, if the vertical variation in the I_{ss} and the soil suction changes (in pF units; Δpf) are both known (or can be estimated) for a soil profile of n layers, then the vertical strains can be calculated in each layer, and integrated to give a nett ground surface movement, according to the equation,

$$y_s = \sum_{i=1}^{n} I_{pt,i} \cdot \Delta p F_i \cdot \Delta z_i = \sum_{i=1}^{n} \alpha \cdot I_{ss,i} \cdot \Delta p F_i \cdot \Delta z_i$$

In equation above, y_s is the nett vertical ground surface movement, provided the suction changes represent the design suction changes for the climatic region or local conditions of the site. In the context of Australian practice, ysis defined as the characteristic ground surface movement at a site, and it is defined in statistical terms as being that ground surface movement that is likely to occur in an undeveloped area, over a nominated time period, largely as a result of climatic influences. The required statistical significance is achieved by selecting an appropriate distribution of likely suction changes with respect to depth. The value of y_s is then used, to estimate mound heights and edge distances that can be used in the structural design of shallow foundations.

It is now routine in Australian practice for a "site classification" to be carried out to guide foundation design, for every new residential structure. This is usually carried out as part of a larger geotechnical investigation that also assesses issues such as slope stability, and that makes specific recommendations on good site development practice. For a single, isolated site, this usually involves logging, and possibly testing, the soils from at least one borehole a minimum of 3 boreholes per site are suggested for regions where the depth of the design suction change extends beyond 3 m and the soil profile in the general area is known to be variable. For a larger subdivision on fairly uniform soil profiles and relatively shallow depths of movements, the number of soil profiles logged, and shrink swell tests performed, may be as few as one for every three to five building sites.

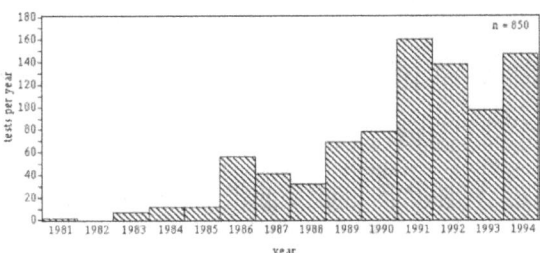

Bearing Capacity

Concept of Bearing Capacity

All civil engineering structures whether they are buildings, dams, bridges etc. are built on soils. A foundation is required to transmit the load of the structure on a large area of soil. The foundation of the structure should be so designed that the soil below does not fail in shear nor there is excessive settlement of the structure. The conventional method of foundation design is based on the concept of bearing capacity.

The bearing capacity of foundation is the maximum load per unit area which the soil can support without failure. It depends upon the shear strength of soil as well as shape, size, depth and type of foundation. Figure shows a typical load vs. settlement curve of a footing. From the figure it is clear that as the footing load is increased, the settlement also increases.

The settlement increases linearly with load at the initial stage. On further increase in load, the settlement increases more rapidly and then continues to increase without any appreciable increase in load. This stage is called failure of foundation i.e., the soil has reached its capacity to bear load.

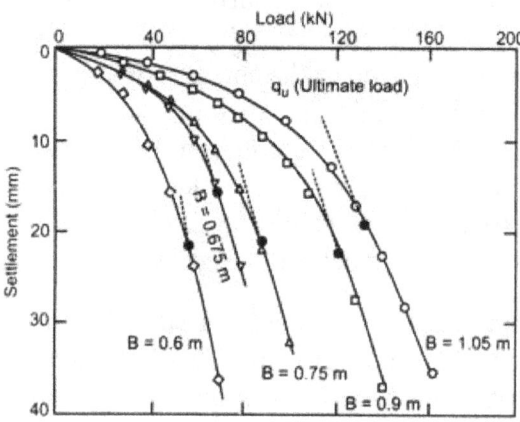

Figure: Load vs settlement curveof footing

To avoid bearing capacity failure of foundation it is essential to take into consideration, before design of foundation, two types of action by the soil when subjected to load:

(i) The bearing capacity should be low enough to ensure that the settlement caused is not excessive.

(ii) The bearing capacity should be such that excessive shear strain is not caused.

Bearing Capacity of Shallow Foundations (Terzaghi Analysis)

Assumptions in Terzaghi's Analysis

1. The footing is strip one at shallow depth and has rough base; (L > 5B, D > B, where L = length, B = width and D = depth of the footing).

2. The soil is homogeneous, isotropic# and relatively incompressible.

3. The failure zones do not extend above the horizontal plane through the base of the footing.

4. The elastic zone has straight boundaries inclined at $\mu = \varphi$ to the horizontal and the plastic zones fully developed.

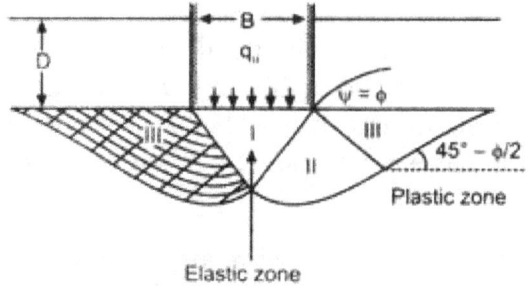

Figure: Failure surface and zones in the Terzaghi's analysis

Also called general bearing capacity equation for strip footing.

$$q_u = CN_c + 0.5\,\gamma BN\gamma + qN_q$$

where q_u = ultimate bearing capacity,

q = overburden pressure at the base,

$\quad$ = γD (use γD, if submerged) C – cohesion of soil

γ = unit weight of soil at base level of foundation

(use γ of submerged)

B = Width of foundation

D = Depth of foundation

N_C, N_g and N_q are bearing capacity factors which depend on φ (angle of internal friction).

Bearing Capacity from Building Codes

For preliminary design of any structure and for design of foundation of lightly loaded structures, the presumptive bearing capacity may be used. Table gives presumptive safe bearing capacities for various types of soil.

S No.	Type of soil	Safe bearing capacity	
1.	Gravel, sand and gravel, com- pact and offering high resistance to penetration when execrated by tools	44.0	440
2.	Coarse sand, compact and dry	44.0	440
3.	Medium sand, compact and dry	24.5	245
4.	fine sand, silt (dry lumps easily pulverized by the fingers)	15.0	150
5.	Loose gravel or sand gravel mixture, loose coarse to medium sand , dry	24.5	245
6.	Find sand , loose and dry (B) cohesive soils	10.0	100
7.	Soft shale hand or stiff clay in deep bed, dry	44.0	440
8.	Medium clay, readily indented with a thumb nail	24.5	245
9.	Moist clay and sand clay mixture which can be indented with strong thumb pressure	15.0	150
10.	Soft clay indented with moderate thumb pressure	10.0	100
11.	Very soft clay which can be penetrated several centimeters with the thumb	5.0	50
12.	Black cotton soil or other shrinkable or expansive clay in dry condition (50% saturation) (C) Peat	To be determined after in-vestigation	
13.	Peat (D) made up ground	-do-	
14	Fills or made up ground	-do-	

Note 1:

Values of bearing capacity listed are from shear consideration only.

Note 2:

Values listed in the table are very much rough for the following reasons:

(i) Effect of depth, width, shape and roughness of foundation have not been considered.

(ii) Effect Of angle of friction, cohesion, water-table, density etc., have not been considered.

(iii) Effect of eccentricity and indication of loads has not been considered.

Note 3:

Dry means that the groundwater level is at a depth not less than the width of foundation below the base of the foundation.

Note 4:

For cohesion less soils, the values listed in the table shall be reduced by 50% if the water table is above or near the base of footing.

Note 5:

Compactness of cohesion less soils may be determined by driving a cone of 65 mm dia and 60° apex angle by a hammer of 65 kg falling from 75 cm If the corrected N-value for 30 cm penetration is less than 10, the soil is called loose, if N lies between 10 and 30, it is medium and if more than 30, the soil is called dense.

Factors Affecting Bearing Capacity of Soils

- Type of soil
- Physical characteristics of foundation
- Soil properties
- Type of foundation
- Water table
- Amount of settlement
- Eccentricity of loading.

The following Factors Affect the Bearing Capacity of Soils

(i) Type of soil:

The bearing capacity of soils depends upon the type of soil. Depending upon the type of soil, the bearing capacity of soil is different which is clear from Terzaghi bearing capacity equation.

$$q_u = CN_C + 0.5\,yBNy + qN_q$$

For purely cohesion less soil

$$C = 0$$

Equation $q_u = CN_C + 0.5\,yBNy + qN_q$ reduce to,

$$q_u = 0.5\,yBNy + qN_q$$

For purely cohesive soil,

$$\varphi = 0,$$

the values of bearing capacity factors are,

$$Nc = 5.7$$

$$Nq = 1\ and\ N\gamma = 0$$

Equation $q_u = CN_C + 0.5\,yBNy + qN_q$ is then

$$q_u = 5.7C + q$$

(ii) Physical characteristics of foundation

Physical characteristics like width, shape and depth of foundation affect the bearing capacity of soils. equation $q_u = CN_C + 0.5\,yBNy + qN_q$ shows that the bearing capacity of soils depends upon the width B and depth (D) of foundation. So any change in the value of B and D of foundation will affect the baring capacity.

The shape of foundation also affects the bearing capacity which is as follows:

For square footings

$$q_u = 1.2\,CNc + 0.4\,\gamma BN\gamma + \gamma DNq$$

For circular footings

$$q_u = 1.2\,CN_C + 0.3\,\gamma BN\gamma + \gamma DN_q$$

where B is the diameter of circular footing.

(iii) Soil properties

Soil properties like shear strength, density, permeability etc., affect the bearing capacity of soil. Dense sand will have more bearing capacity than loose sand as unit weight of dense sand is more than loose sand.

(iv) Type of foundation

The type of foundation selected also affects the bearing capacity of soils. Raft or mat foundation adopted supports the load of structure safely by spreading the load to a wider area, even if the soil is having low bearing capacity.

(v) Water table

When the water is above the base of the footing, the submerged unit weight of soil is used to calculate the overburden pressure and the bearing capacity of the soil reduces by 50%.

For any position of the water table the general bearing capacity may be modified as under

$$q_u = CN_c + 0.5\gamma BN_\gamma R_w + \gamma DN_q R'_w$$

where R_w and R'_w are the reduction factor for water table

$$R_w = \frac{1}{2}\left(1 + \frac{D_w}{B}\right)$$

and $\quad R'_w = \frac{1}{2}\left(1 + \frac{D'_w}{B}\right)$

At $D_w = 0, R_w = 0.5, R'_w = 1$

$D_w = B, R_w = \frac{1}{2}\left(1 + \frac{1}{1}\right) = 1$

$R'_w = 1$

At $D'_w = 0, R_w = 0.5, R_w = 1$

At $D'_w = D, R'_w = 1, R_w = 0.5$

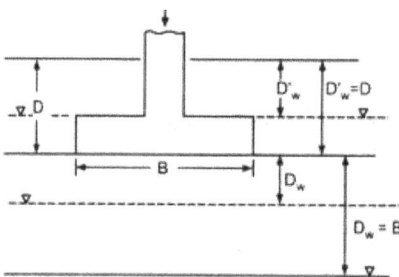
Figure: Effect of water table

(vi) Amount of settlement

The amount of settlement of the structure also affects the bearing capacity of soil. If the settlement exceeds the possible settlement, the bearing capacity of soil is reduced.

(vii) Eccentricity of loading

If the load acts eccentrically in a footing the width 'B' and length 'L' should be reduced as under,

$B' = B - 2e$

$L' = L - 2e$ and

$A' = B' \times L$

The ultimate bearing capacity (qu) of such footings are determined by using B' and L' instead of 8 and L. Hence qu is less than that corresponds to actual size of footing as shown in figure below.

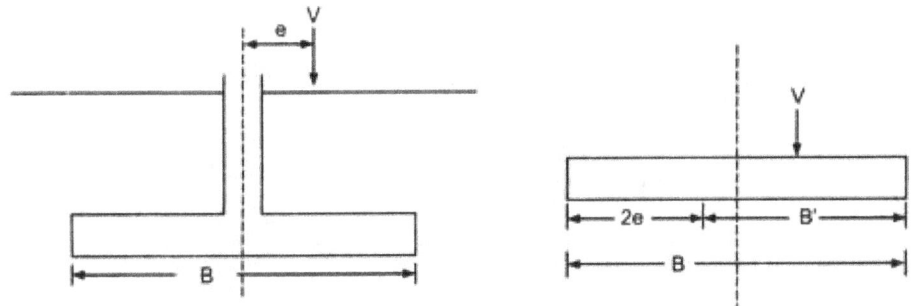

Figure: Effective width for a footing subjeccted to eccentric loading

Concept of Vertical Stress Distribution in Soils Due to Foundation Loads

When a soil mass is loaded, vertical stresses are developed in the soil. The estimation of vertical stresses at any point in a soil mass due to external loading are of great significance in the prediction of settlement of buildings, bridges, embankments and other structures. The stresses due to external loading are greatest at shallow depths, close to the point load application and they become smaller as the vertical distance below the load or the horizontal distance from the load increases.

The vertical stress distribution in a soil mass depends upon:

(i) The nature of loading i.e., the manner of load placement, load distribution and shape of the loaded area,

(ii) Physical properties of soil like Poisson's ratio, modulus of elasticity, compressibility etc.

In determining the stresses below a foundation, it is generally assumed that the soil behaves as an elastic medium with identical properties at all points and in all directions. Many formulae based on theory of elasticity have been used to compute stresses in soils. One such formula was first developed by Boussinesq for the stresses and deformation in the interior of a soil mass due to vertical point load. A British scientist Westergaard in 1938 also proposed a formula for computation of vertical stress in the soil mass due to vertical point load.

Point Load

Business's Formula

Business's formula is based on the following assumptions

(i) The soil mass is linearly elastic, homogeneous, isotropic and semi- infinite.

(ii) The load acts as a vertical concentrated load.

(iii) The soil is weightless.

The equation for vertical stress at a point as shown in figure below.

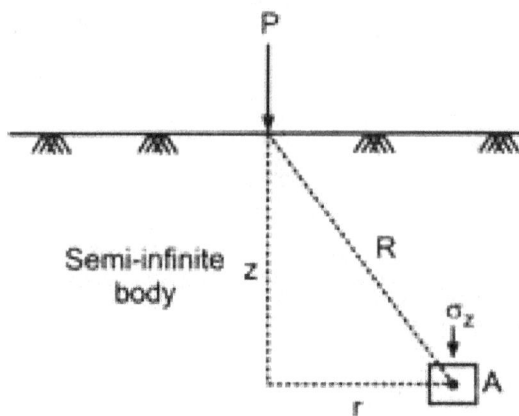

Figure:Vertical stress within an soil mass.

The equation for vertical stress at a point as shows in figure above,

$$\sigma_z = \frac{P}{Z^2} K$$

Where K=a dimensional factor

$$\left[1 + \left(\frac{r}{z}\right)\right]^{5/2}$$

P= Point load

Z= vertical depth of point A form ground level

r = Horizontal distance from A to the vertical axis through the point load P.

S. No.	$\frac{r}{z}$	K	S. No	$\frac{r}{z}$	K
1.	0	0.4775	6.	0.6	0.2214
2.	0.1	0.4657	7.	0.7	0.1762
3.	0.2	0.4329	8.	0.8	0.1386
4.	0.3	0.3849	9.	0.9	0.1083
5.	0.4	0.3294	10.	1.0	0.0844
			11.	2.0	0.0085

Line Load

The equation for vertical stress due to a line load P1 per unit length on the surface at a point located at a depth z and distance x laterally as shown in figure below is

$$\sigma_z = 2p_1 /$$

$$z^3 / (x^2 + z^2)^2$$

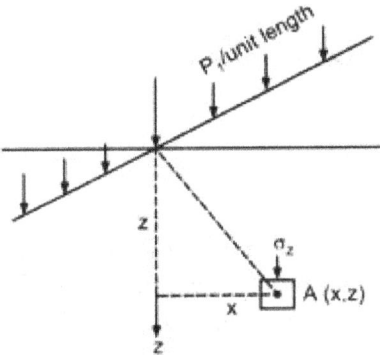

Figure:Vertical stress due to line land.

Uniformly Loaded Strip

The equation for vertical stress due to a uniform load q on a strip area of width B and infinite length in terms of σ and θ as shown in the figure below is,

$$\sigma_z = q / \pi \; (\alpha + Sin \; \alpha Cos \; 2\theta)$$

Below the center of the strip, vertical stress o at a depth z is given by,

$$\sigma_z = q / \pi (a + sin \; \alpha) \; (\theta \text{ is zero and } cos 2\theta = 1)$$

or $\sigma_2 = ql \; oz$

The values of the influence factor are given in table below,

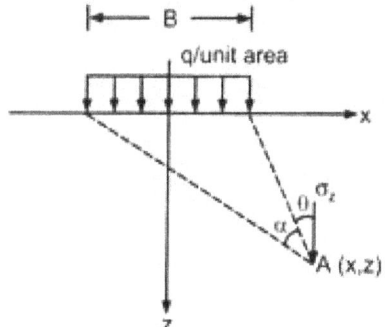

Figure:Vertical stress due to uniformly loaded strip

S.No	$\dfrac{r}{z}$	I_{az}	S.No	$\dfrac{Z}{B}$	I_{az}	S No.	$\dfrac{Z}{B}$	I_{az}
1.	0.2	0.978	7.	1.4	0.420	13.	2.8	0.223
2.	0.4	0.880	8.	1.8	0.336	14.	3.0	0.209
3.	0.6	0.755	9.	2.0	0.306	15.	4.0	0.160
4.	0.8	0.642	10.	2.2	0.280	16.	5.0	.0130
5.	1.0	0.551	11.	2.4	0.270	17.	6.0	0.110
6.	1.2	0.477	12.	2.6	0.240			

Soil Properties Governing Choice of Foundation Type

The following Properties of Soil Govern the Choice of Foundation Type

(i) Bearing capacity of soil

(ii) Settlement of soil

The knowledge of bearing capacity and settlement of the soil is very essential for design of foundation of any structure. The foundation of any structure should be so selected that the soil below does not fail in shear and settlement is within the permissible limits.

If the bearing capacity of soil at shallow depth is sufficient to safely take the load of the structure, a shallow foundation is provided. Isolated footing, combined footing or strip footing are the option for shallow foundation. Deep foundations are provided when soil immediately below the structure does not have adequate bearing capacity. Pile, piers or well are the options for deep foundations. Mat or raft foundations are useful for soil which are subjected to differential settlement or where there is a wide variation in loading between adjacent columns. Table below gives suitability of foundation for buildings based on soil type.

S NO.	soft silty clayss.	soft silty clayss.
1.	Sands and gravels or sands and gravels with small proportions of clay, clays. Soft silty clays	Shallow, strip or spread footings
2.	Soft clays, silty clays.	Strip footings upto 1 m wide or rafts.
3.	Peat, fill	Board piles or driven piles
4.	Deep deposit of loose sand	Rafts, cast-in-situ piles. Driven piles could be used and would density sand.

In-situ Tests for Determination of Ultimate Bearing Capacity

The following in-situ tests may be used to determine the ultimate bearing capacity or allowable bearing capacity of soil:

- Plate load tests
- Standard penetration test
- Dynamic cone penetration test
- Static cone penetration test
- Pressure meter test

Plate Load Test

Plate load test essentially consists of loading a rigid plate at the foundation level and recording the settlements corresponding to each load increment. The ultimate bearing capacity is then taken as the load at which the plate starts sinking at a rapid rate. The minimum and maximum recommended sizes to test plate are 30 cm square and 75 cm square respectively. The thickness of the steel plate should not be less than 25 mm. Alam Singh has recommended the size of the test plate to be 32 cm square.

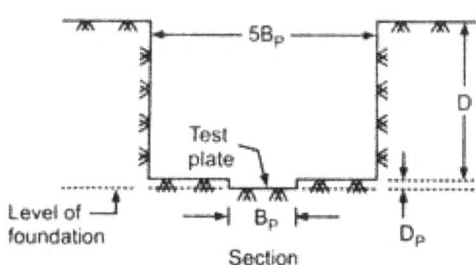
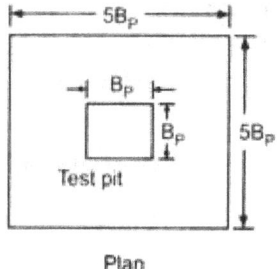

Figure: Test pit

The test is carried in a pit having width equal to 5 times the width of the test plate. At the centre of the pit, a small square hole is dug whose size is equal to the size of the plate and the bottom level of the pit correspond to the level of the actual foundation.

The Loading to the test Plate may be Applied by the following Two Methods

(a) Gravity Loading Platform method

(b) Reaction Truss Method

Reaction truss loading is found convenient and less time consuming, hence generally used. For this purpose, a steel truss is anchored to ground across the pit. A hydraulic jack with attached pressure gauge is placed between the underside of the truss and the test plate. At least two dial gauge, having accuracy of 0.2 mm, is used to measure the settlement of the test plate. The dial gauges are mounted on independent datum bar and are just touching the test plate.

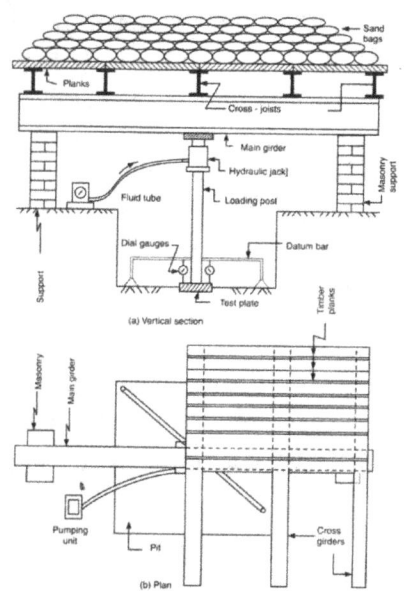
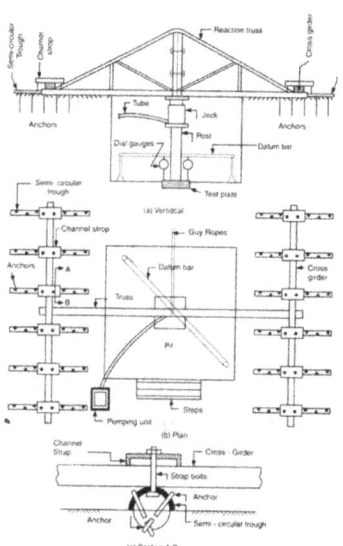

Figure: Plate load test (reaction by gravity loading)   Figure: Plate load test (reaction by truss)

Before starting the test, a seating pressure of 70 gm/cm2 is applied to the plate (as recommended by IS 1888-1962). It is then removed and dial gauges are set to zero reading. Load is then applied in cumulative equal increments; say about 1/5 of the expected safe bearing capacity or 1/10 of the expected allowable bearing capacity. Settlement should be recorded for each increment of load after an interval of 1, 4, 10, 20, 40 and 60 min and thereafter at hourly intervals, until the rate of settlement becomes less than about 0.02 mm per hour. After this, load is increased to the next higher value and the process is repeated.

Testing in continued until one of the following stages is attended

(a) Settlement is at a faster rate indicating shear failure.

(b) The applied pressure exceeds 3 times the proposed allowable bearing pressure.

(c) The total settlement exceeds 10 percent the width of the test plate. The load is then released. If desired, rebound observation may be taken.

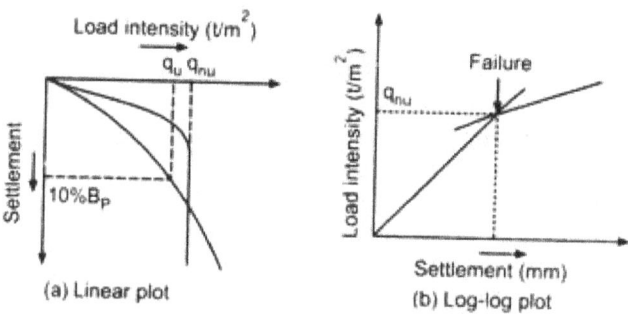

Figure: Load settlement curve

Interpretation

The load intensity and settlement observations of the test are plotted, as shown in figure above, in the linear scale as well as log-log scale. IS 1888-1962, recommends a log-log plot giving two straight lines the intersection of which may be considered the failure of the soil. When the failure point is not clear in the graph, failure may be assumed at a settlement of 10% of plate width. The load intensity corresponding to the failure point gives the ultimate bearing capacity and a factor of safety of 2.5 or 3 on ultimate bearing capacity may be used to get safe bearing capacity of soil.

Effect of Size of the Plate on Bearing Capacity

Bearing capacity of sands and gravels increases with the size of the footing. The bearing capacity obtained from plate load test for sandy soils will be different from the actual bearing capacity of foundation as the size of foundation will be more than that of plate. For all practical purpose the plate load test data is extrapolated to get the bearing capacity of actual footing.

For sandy soils:

$$q_{uf} = q_{up} \times B_F / B_P$$

where

q_{uf} = ultimate bearing capacity of actual footing

q_{up} = ultimate bearing capacity from plate load test

B_f = width of footing

B_p = width of plate

For clay soils

$$q_{uF} = q_{up}$$

Effect of Size of Plate on Settlement

The settlement of footing varies with its size. So the settlement obtained from plate load test may not be the same as that of actual footing.

Following Relationship are used to find out Settlement of Actual Footing

For clay soils

$$S_F = S_P \times B_F / B_P$$

S_p = Settlement of actual footing in mm

S_p = Settlement from plate load test in

B_F Width of footing in meters

B_P Width of plate in meters

For sandy soils

$$S_F = S_P \left[B_F \left(Bp + 0.3 \right) / B_p \left(B_F + 0.3 \right) \right]^{-2}$$

Limitations

Plate load test data reflect the characteristics of soil only within a depth equal to twice the width of plate. Since the actual foundation is larger than the size of plate, the plate load test does not truly represent the actual soil condition in case of a non-homogeneous soil as shown in figure below.

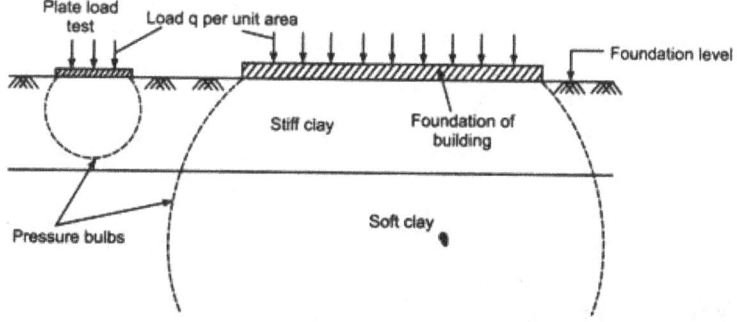

Figure: Plate load test in non-homogeneous soil

Plate load test is essentially a short duration test (run in few hours), So no indication of long term consolidation settlement in clays is obtained.

This test should not be relied on to obtain the ultimate bearing capacity of Sandy soils as the scale effect gives very misleading results.

The proximity of water table may be within the influence of the footing and not that of the test plate, as the effect of submergence is to reduce the bearing capacity of granular soils by 50%.

Bearing Capacity Based on Standard Penetration Test (SPT)

In case of cohesionless soils, the SPT results are used to determine the ultimate bearing capacity of soils by the following methods:

By using chart given by Peck, Hanson and Thornburn:

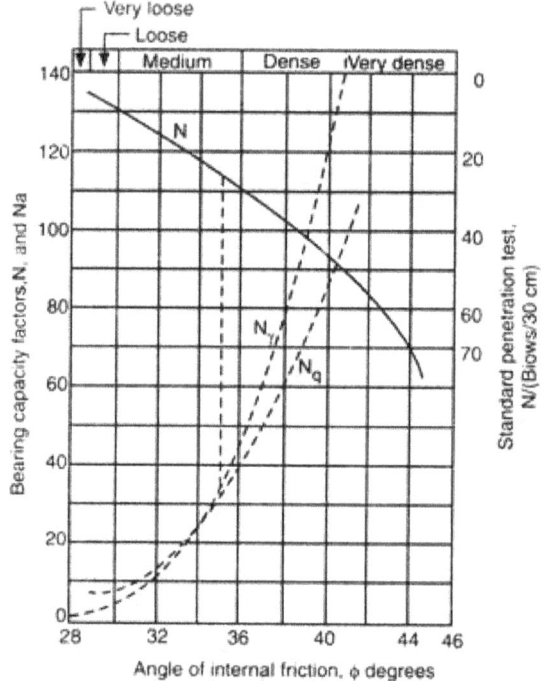

Figure: Bearing capacity of footing bassed on N-value

Figure above shows variation of bearing capacity factors N_q and Ng w.r.t. ϕ as well as corrected N-value.

This chart can be used directly for N_q and $N\gamma$ for use in bearing capacity equation written below

$$q_u = CN_c + qN_q + 0.5gBNg$$

For cohesionless soil

$C = 0$ and the above equation reduces to

$q_u = qN_q + 0.5\,yBNy$

N_q and Ny values are obtained directly from figure above.

Overconsolidation Ratio

Overconsolidation ratio (OCR) is used to indicate the stress history of a soil deposit. Value of OCR can be determined by conducting an odometer test using an undisturbed soil sample. However, retrieving undisturbed soil samples from boreholes is costly, and therefore, the number of boreholes that can be drilled and number of undisturbed soil samples that can be retrieved are limited. Moreover, conducting a laboratory test is time consuming, and the test results can be influenced by several factors such as sample disturbance and interpretation method employed for determining the maximum past consolidation pressure. In case of geotechnical projects, particularly underground construction projects, such as excavation and tunneling, it is always desirable to have more detailed information about the underground. The piezocone sounding (uCPT) is an economic and efficient in situ testing method that can be conducted with a pitch of a few tens of meters to identify the local variations in the soils and their properties, so that the results can be used to select suitable construction methods as well as to achieve a detailed design. uCPT provides nearly continuous measurements of the tip resistance (q_t), sleeve friction (f_s), and pore water pressure (u_2) at the shoulder of the cone (standard cone). From the results of uCPT, soil profiles and other engineering properties of the subsoil, such as undrained shear strength (s_u) and OCR can be estimated.

Methods for Evaluating OCR

Existing methods

Method 1: Using undrained shear strength (s_u)

Piezocone sounding results are widely used for estimating the undrained shear strength (s_u) of clayey soils. The basic concept is to equate the values of s_u obtained from the piezocone sounding results to the value of su derived from other empirical or theoretical methods that involve the OCR. From the piezocone sounding results, s_u can be evaluated by equation below.

$$s_u = \frac{(q_t - \sigma_{v0})}{N_{kt}}$$

where qt is the corrected cone tip resistance, N_{kt} is a cone bearing factor, and σ_{v0} is the total stress in the vertical direction. Table lists some proposed values (ranges) of N_{kt}. Ladd and Foott found that the relationship between the S_v / σ'_{v0} (σ'_{v0} is the effective stress in the vertical direction) ratio and OCR can be expressed by the following empirical equation:

$$\frac{S_v}{\sigma'_{v0}} = S \cdot OCR^m$$

where S and m are constants. Ladd and DeGroot suggested that for most soils, m is approximately 0.8. Parameter S varies with the site. Ladd and DeGroot recommended S = 0.22 for most homogeneous inorganic clays. For organic soils, not including peats, they suggested S = 0.25. Using the results of a simple shear (DSS) test with normal consolidated soil samples, Wroth proposed an equation for estimating S as follows:

$$S = \frac{1}{2} \mathrm{Sin}\, \phi'$$

where ϕ' is the internal friction angle of a soil. By combining equations. $s_u = \frac{(q_t - \sigma_{v0})}{N_{kt}}$, and

$\dfrac{S_v}{\sigma'_{v0}} = S \cdot OCR^m$, an expression for the OCR can be derived to be as follows,

$$OCR = \left[\left(\frac{q_t - \sigma_{v0}}{N_{kt}} \right) \cdot \frac{1}{\sigma'_{v0} \cdot S} \right]^{1/m}$$

Method: Using tip resistance (Q_t)

Another empirical method that has been extensively used for evaluating the OCR from uCPT results uses the dimensionless tip resistance (Q_t)

$$Q_t = \left(\frac{q_t - \sigma_{v0}}{\sigma'_{v0}} \right)$$

The relationship between the OCR and (Q_t) can be expressed as

$$OCR = k \cdot Q_t$$

where k is a constant.

Based on the SHANSEP (Stress History and Normalized Soil Engineering Properties) approach, Saye, Lutenegger, Santos, and Kumm proposed a two-parameter method to estimate the value of the OCR from Q_t as follows:

$$OCR = \left(\frac{Q_t}{Q_{NC}} \right)^{1/mCPTu}$$

where Q_{NC} and $mCPTu$ are two empirical constants. It was suggested that Q_{NC} can be estimated from the plasticity index (Ip) and $mCPTu$ from the liquid limit (w_L). Equation $OCR = k \cdot Q_t$ only needs one empirical parameter, but equation $OCR = \left(\dfrac{Q_t}{Q_{NC}} \right)^{1/mCPTu}$ requires two empirical parameters; therefore, equation $OCR = k \cdot Q_t$ is considered more practical and is used in this study.

Method: Based on cavity expansion and critical state soil mechanics theories

Mayne proposed a method for estimating the OCR by combining the theories of cavity expansion and critical state soil mechanics. The OCR can be computed by the following equation:

$$OCR = 2 \left[\frac{1}{1.95M + 1} \left(\frac{q_t - u_2}{\sigma'_{v0}} \right) \right]^{1/\Lambda}$$

where M is the slope of the critical state line in a p'–q plot (where p' is the mean effective stress and q is the deviator stress) defined as, $M = \dfrac{6 \sin\phi'}{3 - \sin\phi'}$. In equation $OCR = 2 \left[\dfrac{1}{1.95M + 1} \left(\dfrac{q_t - u_2}{\sigma'_{v0}} \right) \right]^{1/\Lambda}$, u_2 is the pore pressure measured at the shoulder of the cone and $\Lambda = 1 - \kappa/\lambda$ (λ and κ are slopes of the virgin loading and unloading–reloading curves in the void ratio, e versus ln (p') plot). The range of κ/λ is $1/5$ to $1/10$, , and therefore, $\Lambda = 0.8 - 0.9$.

Trevor and Mayne modified equation $OCR = 2 \left[\dfrac{1}{1.95M + 1} \left(\dfrac{q_t - u_2}{\sigma'_{v0}} \right) \right]^{1/\Lambda}$ as follows:

$$OCR = 2\left(0.029 + 0.409M\right)\left[\frac{1}{1.95M + 1}\left(\frac{q_t - \upsilon_2}{\sigma'_{\upsilon 0}}\right)\right]^{1/\Lambda}$$

Proposed Method

The modified Cam Clay (MCC) model is one of the most widely used models for soft clay soils. Under a triaxial compression condition, MCC predicts the undrained shear strength (s_u) of a soil sample as,

$$\left(s_u\right) = \frac{p'}{2^{1 + \Lambda}}M\left(\frac{M^2 + \eta^2}{M^2}\right)^{\Lambda}\left(OCR\right)^{\Lambda}$$

Where $\eta = q / p', p' = \sigma'_{\upsilon 0}(1 + 2K_0)/3, and\ q = (1 - K_0)\sigma'_{\upsilon 0}$. K_0 is the coefficient of earth pressure at-rest that is a function of the OCR, and can be expressed by the equation of Mayne and Kulhawy as,

$$K_0 = \left(1 - Sin\phi'\right)OCR^{sin\phi'}$$

By combining equations $s_u = \frac{(q_t - \sigma_{\upsilon 0})}{N_{kt}}$ and (s_u) $s_\upsilon = \frac{p'}{2^{1 + \Lambda}}M\left(\frac{M^2 + \eta^2}{M^2}\right)^{\Lambda}\left(OCR\right)^{\Lambda}$, an expression for the OCR can be derived as followed:

$$OCR = \left(\frac{q_t - \sigma_{\upsilon 0}}{N_{kt}}\right)^{1/\Lambda}\left(\frac{2^{1+\Lambda}}{MP'}\right)^{1/\Lambda}\left(\frac{M^2}{M^2 + \eta^2}\right)$$

In equation $OCR = \left(\frac{q_t - \sigma_{\upsilon 0}}{N_{kt}}\right)^{1/\Lambda}\left(\frac{2^{1+\Lambda}}{MP'}\right)^{1/\Lambda}\left(\frac{M^2}{M^2 + \eta^2}\right)$, η is also a function of the OCR; there-fore, to predict the OCR from the same equation, an iterative approach is needed. In this approach, we begin by assuming the OCR = 1.0 and input it to the right side of equation

$$OCR = \left(\frac{q_t - \sigma_{\upsilon 0}}{N_{kt}}\right)^{1/\Lambda}\left(\frac{2^{1+\Lambda}}{MP'}\right)^{1/\Lambda}\left(\frac{M^2}{M^2 + \eta^2}\right)$$, to obtain a new value of the OCR on the left side. Then, the newly obtained value of the OCR is input into the right side of the equation, and the procedure is repeated until the difference between the inputted value of the OCR on the right side and the resulting value of the OCR (estimated value) on the left side is less than 0.05. For the value of Λ if there is no measured value available, $\Lambda = 0.85$ is suggested.

Soil Gradation

The behavior of soil under external loads depends mainly on its particle size and arrangement of particles. It is therefore very important to study the size, shape and gradation of soil particles. Soil is classified on the basis of the size of their particles. The purpose of soil classification is to arrange various types of soils into groups according to their engineering properties.

Particle Size

Individual solid particle in a soil can have different sizes and this characteristic of soil can have a significant effect on its engineering properties. Size of particles that constitute soils may vary from boulders to that of large molecules.

The soil particles coarser than 0.075 mm constitute the coarse fraction of soils. Particles finer than 0.075 constitute the finer fraction of soils. Coarse fractions of soil consist of gravel and sand. Silt and clay are the fine fractions of soils.

Soil is classified based on the particle size. There are various particle size classifications in use.

A few of these classification systems are given below.

U.S. Bureau of Soil Classification System

Figure below gives the particle sizes and corresponding soil types according to this classification.

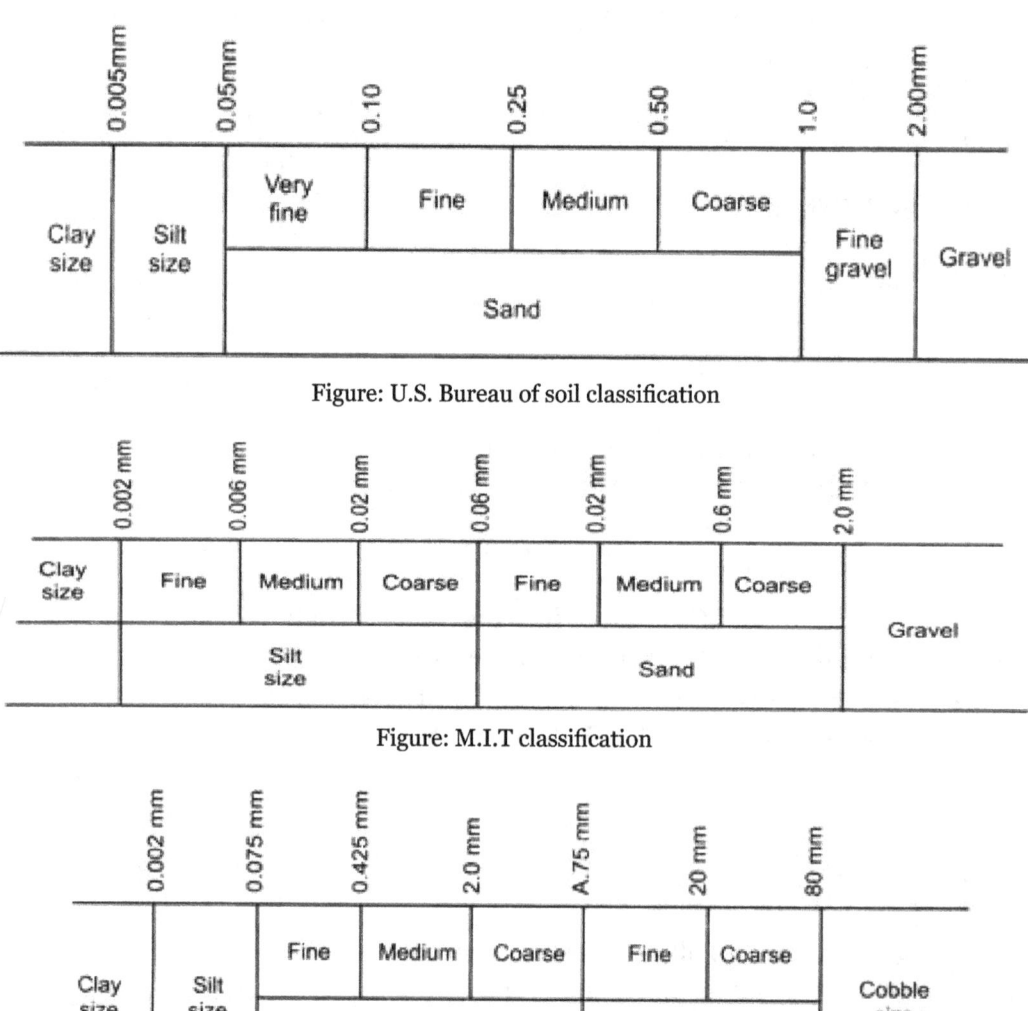

Figure: U.S. Bureau of soil classification

Figure: M.I.T classification

Figure: BIS soil classification

Particle Shape

Shape of the particles helps in determining the property of soil. The shape of particles varies from very angular to well round. Angular particles are generally found near the rock from which they are formed. Angular particles have greater shear strength than rounded ones because it is more difficult to make them slide over one another.

Depending upon the ratio of length, width and thickness, the particles are classified as:

(i) Bulky particles

When the length, width and thickness of particles are of same order of magnitude, the particles are called bulky. Cohesion less soils has bulky particles.

Bulky particles are further classified as

Angular, sub-angular, sub-rounded, rounded and well rounded,

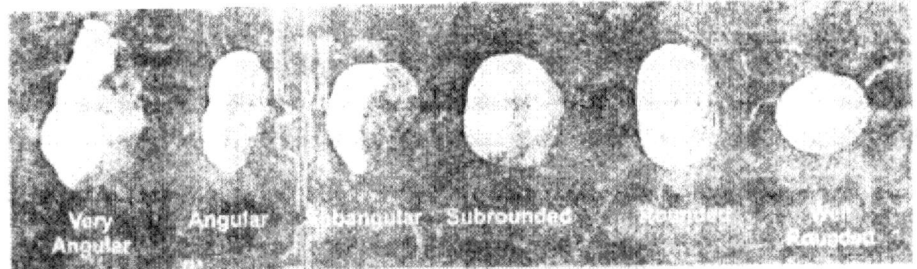

Figure: Shapes of particles

    (a) Plate like flaky

    (b) Elongated (needle like)

(ii) Flaky particles

Flaky particles are also called plate like particles. These particles are mostly present in cohesive soils and are extremely thin as compared to its length and breadth. Figure below shows flaky particle.

(iii) Elongated particles

Elongated soil particles are just like hollow rods. It is a special type of particles and are available in clay minerals i.e., Halloy site, peat, asbestos etc. Figure below shows elongated particles.

(a) Plate like flaky   (b) Elongated (needle like)

Figure: Clay particles

Effect of Shape on Engineering Properties

Engineering properties of soils are affected by the shapes of particles. Angular particles have greater shear strength than rounded ones because it resists displacement. Angular particles have move tendency of fracture. Coarse grained soils have bulky particles.

These soils can support heavy loads in static condition. Settlement of such soils is more when subjected to vibration. Flaky particles are highly compressible and so clay soil which contained these particles is highly compressible. These soil particles deformed easily under static load. Clay soils are more stable when subjected to vibration.

Gradation of Soil

Gradation describes the distribution of different sizes of individual particles within a soil sample. The particle size distribution curve is used to define the grading of soil.

A soil sample may be either:

- Well graded
- Poorly graded
- Gap graded

(a) Well graded

A soil sample is said to be well graded if it has all sizes of materials present in it.

(b) Poorly graded

Poorly graded soil is a soil sample in which most of the particles are approximately of the same size.

(c) Gap graded

A soil sample is said to be gap graded if at least one particle size is completely missing in it. Gap graded soils are sometimes considered a type of poorly graded soil.

Influence of Gradation on Engineering Properties of Soils

Gradation of soils affects the engineering properties like shear strength, compressibility, etc. Well graded soils have more interlocking between the particles and thus a higher friction angle, than those that are poorly graded. The compressibility of well graded soils is almost none and that of poorly graded soils are more than that of well graded soil. Hence permeability of poorly graded soil will be more than that of well graded soil. Well graded soils are more suitable for construction than the poorly graded soils.

Particle Size Distribution Curve

It is also known as gradation curve and represents the distribution of particles of different sizes in the soil sample. It is a graph of the results obtained from sieve analysis, on a sami-log paper with percentage finer on the arithmetic scale as ordinate and the particle size as abscissa on log scale. Figure shows particle size distribution curve. Curves on the left side of the graph, such as soil A, indicate fine grained soils, while those on the right of the curve, such as soil B, indicate coarse grained soils.

Steep curves, such as soil C indicate soil with a narrow range of particle sizes i.e., poorly graded soils. Flat curves, such as soil D, contains a wide range of particle sizes i.e., well graded soils. Curves, in which nearly flat-zones are observed, such as soil E, are of gap graded soils. The particle diameters that correspond to certain percent passing values for a given soil are known as the D-sizes. For example D10 represents a size such that 10% of the particles are finer than this size.

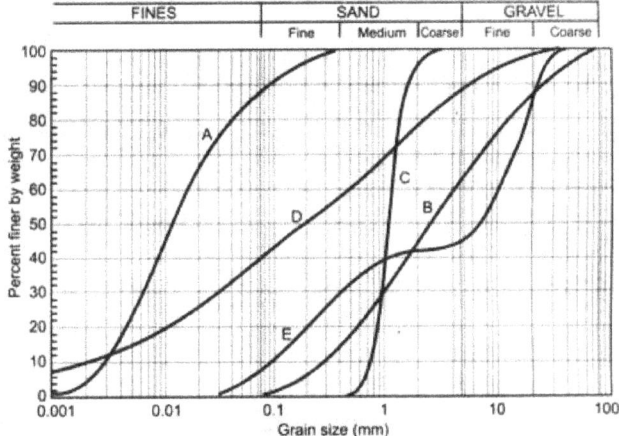

Figure: Particle size distrbution curve

Coefficient of uniformity, Cu and the coefficient of curvature, Cc, are the parameters based on D-size to define grading. Coefficient of uniformity, and coefficient of curvature,

Where D

$$Cu = D_{60} / D_{10}$$

$$Cc = (D_{30})^2 / D_{10} \times D_{60}$$

Where,

D_{10} — Particle diameter at which 10% of the soil mass is finer than this size,

D_{30} — Particle diameter at which 30% of the soil mass is finer than,

D_{60} — Particle diameter at which 60% of the soil mass is finer than this size.

well graded soils have high C_u values and poorly graded soils have low C_u, values. If all the particles of soil mass are of the same size C_u is unity.

C_c lies between 1 to 3 for well graded soil.

$C_u > 6$ *for samples*

$C_u > 6$ *for samples*

The gradation of soil is determined by the following criteria:

Uniform soil: $Cu = 1$

Poorly graded soil: $1 < Cu < 4$

Well graded soil: $Cu > 4$

Sieve Analysis

It is a laboratory test that measures the particle size distribution of a soil by passing it through a series of sieves. The complete sieve analysis is divided into two parts—coarse analysis and fine analysis.

The entire soil sample is divided into two fractions by sieving it through 4.75 mm IS sieve. Soil retained on it is termed as gravel fraction and is kept for the coarse analysis. Soil passing 4.75 mm sieve is used for fine sieve analysis.

For coarse sieve analysis IS: 100, 63, 20, 10 and 4.75 mm sieves are used.

For fine sieve analysis IS: 2.0 mm, 1.0 mm, 600, 425, 300, 212, 150 and 75 micron sieves are used.

Sieve analysis is performed by arranging the set of sieves in order i.e., by keeping the largest aperture sieve at top and smallest aperture at the bottom. A lid is placed at the top sieve and a pan at the bottom sieve.

Dry Sieving

Figure: Set of sieves

Soil sample is placed at the top sieve and is covered with the lid. The entire set of sieves is then placed in a sieve shaker. After 10 to 15 minutes of shaking in the sieve shaker, sieves are removed from the shaker. Soil sample retained on each sieve is weighed. Percentage of soil retained in each sieve is calculated and finally percentage passing through each sieve is obtained. Table below shows the specimen calculation sheet.

Figure: Sieve shaker

Figure: Sieve showing the opening

Wet Sieving

Wet sieving is advisable for soil sample passing through 4.75 mm sieve. The soil sample passing 4.75 mm sieve is taken in a tray and is covered with water. 2gms of sodium hexametaphosphate per liter of water used is then added to the soil. The mixture is thoroughly stirred and left for soaking.

The soaked soil sample is washed on 75 micron-sieve until the water passing the sieve is clear. The soil retained on 75 micron sieve is taken on a tray and dried. The dry soil is then sieved through set of sieve used for fine grained sieving. Percentage retained and percentage passing through each sieve is calculated.

Fine-grain analysis is carried out by the hydrometer method.

Table: Calculation sheet for sieve analysis Weight of dry sample—1000 gm:

S No	IS Sieve	Partical size (D) in mm	Weight re-tained (gm)	Percentage retained	Commulative Retained	% finer (N)
1.	100 mm	100 mm				100
2.	63 mm	63 mm	2	0.2	0.2	99.8
3.	20 mm	20 mm	40	4.0	4.2	95.8
4.	10 mm	10 mm	33	3.3	6.5	93.5
5.	4.75 mm	4.75mm	49	4.9	11.4	88.6
6.	2 mm	2 mm	87	8.7	20.1	79.9
7.	1 mm	1mm	96	9.6	29.7	70.3
8.	600 micron	0.6 mm	140	14.0	43.7	56.3
9.	425 micron	0.425 mm	170	17.0	60.7	39.3
10.	300 micron	0.300 mm	93	9.3	70.7	30.0
11.	212 micron	0.212 mm	85	8.5	78.5	21.5
12.	150 micron	0.150 mm	33	3.3	81.8	18.2
13.	75 micron	0.075 mm	56	5.6	87.4	12.6
14.	Pan		116			
			1000 gm			

Field Identification of Soils

In field identification of soil, the engineer concerned first determines whether the soil is coarse grained or fine grained. To make this determination, soil sample is spread on a flat surface. If more than half of the particles are visible to the naked eye, then it is classified as coarse grained or otherwise it is classified as fine grained. If the soil is coarse gained, follow the procedures outlined under the heading coarse grained soil.

Coarse Grained Soil

Once the soil has been determined as coarse grained, further examination is required to determine the grain size distribution, the grain shape and gradation of coarse grained soils. Coarse grained soil is classified as cobble or sand depending on whether more than half of the coarse fraction is of cobble size (76 mm or larger) or sand size (5 mm to 0.074 mm). Soil particles can also be described according to a characteristic shape.

Particle shape may vary from angular to round to flat or elongated. Coarse grained soil may be described as well graded, poorly graded or gap graded. A soil is said to well grade if it has a good representation of all grain sizes. If the soil grains are approximately of same size, then the sample is described as poorly graded. A soil is said to be gap graded if the intermediate grain sizes are absent. The appropriate descriptive terms are listed in the tables.

Table: Soil types and Particle Sizes

Soil type	Description	Size	Familiar example
Coarse grained soil	Cobble	76 mm or large	Size of grape or orange
	Coarse grand	76mm to 19 mm	Walnut or grape
	fire gravel	19mm to 5 mm	Pea
	coarse sand	5 mm to 2mm	Rock salt
	medium sand	2mm to 4mm	Openings of a window screen
	Fire sand	0.4 mm to 0.74 mm	Sugar or table salt
Fire grained soil	Silt or clay	Microscopic or submicro-scopic	
Organic	Peat or muck		Decaying vegetable matter

Description	example
Angular	Irregular with sharp edges such as freshly broken rock
Sub-angular	Irregular with smooth edges
Subrounded	Irregular but smooth as a lump of molding clay
Rounded	Egg or mable shaped, very smooth
Flaky	Flake of mica or sheet of paper
Flat	Ratio of width to thickness greater than 3
Elongated	Ratio of length to width greater than 3

Gradation of coarse grained soil

Description	Meaning
Well graded	All sizes of grains are present
Poorly graded	Single size grains are present
Gap graded	Intermediate grains sizes are absents

Density of coarsed grained soil

Description	N- value	Field test
Very loose	Less than 4	
Loose	4-10	12 mm dia rod can be easily pushed into the soil by hand
Medium dense	10-30	12 mm dia rod can be penetrated into the soil by 2-3 kg hammer.
Dense	30-50	12 mm dia rod can be penetrated 30 mm into the soil by 2-3 kg hammer
Very dense	Grater than 50	12 mm dia rod can be penetrated only a few centimeters into the soil by 2-3 kg hammer.

Fine Grained Soil

Following field tests are performed to classify fine grained soil or for the fine fraction of coarse grained soil

Dilatancy Test

Prepare a part of moist soil having a volume equivalent to a 25 mm cube by adding enough water to make the soil soft but not-sticky. Place the pat in the open palm of one hand and shake horizontally by striking against the other hand several times. If the reaction is positive, water appears on the surface of the pat giving a glossy appearance. On squeezing the sample between the fingers the water and glossiness disappear from the surface, the soil becomes stiff and cracks.

The phenomenon of appearance of water on the surface of soil on shaking and disappearance on squeezing, followed by cracking is called as "dilatancy". The rapidity of appearance and disappearance of water from the surface of the soil help to identify the character of fines in the soil. Table shows character of fines in soil w.r.t. the positive reactions.

Table: Dilatancy of fine soil

Positive reactions	Type of soil
Quickest	Very fine clean sand
Moderate	Inorganic silt
No reaction	Plastic clay

Dry Strength Test

Prepare a part of soil to the consistency of putty by adding water. Allow the pat to dry by oven, sun or air. The strength is tested by breaking and crumbling the dry pat between the fingers. Dry strength of soil increases with increasing plasticity. Clays have high dry strength and silts have slight dry strength.

Toughness Test

Take a part of soil to the consistency of putty, add water or allow drying as necessary. Roll the soil between the palms into a thread of 3 mm diameter. Fold the thread of soil and repeat the procedure a number of times till the thread starts crumbling when rolled into 3 mm diameter. The crumbled pieces are lumped together and subjected to kneading until the lump crumbles. The threads are stiffer and lumps are tougher at plastic limit for soils having higher clay contents.

Dispersion Test

Pour small quantity of soil in a jar of water. Shake the jar containing soil and water and allow the soil to settle. The coarser particles settle first followed by finer ones. Sands settle in about 30 to 60 secs, silts settle in 30 to 60 mins and clay particles remains in suspension for at least several hours.

Bite Lest

Take a pinch of soil and place between the teeth and grind lightly. Fine sand is felt gritty. Silt have rough feeling but do not stick to the teeth, clays have smooth feeling and stick to the teeth.

Color and Odour Test

Organic soils have darker colors like dark grey, dark brown etc. and a musty odour. The odour can be more noticeable by heating a wet sample. Inorganic soils have clean, bright colors like light grey, brown, red, yellow or white.

Consistency and Plasticity

Consistency

Consistency is a term used to describe the physical states of soil i.e. the degree of coherence between particles of a soil at given water content. Consistency is directly related to water content of soil, but it has been found that at the same water content different soils may have different consistency.

Plasticity

It is the ability of soil to change shape on application of load and to retain the new shape after removal of the load. Fine soil particles like clays exhibit plastic behavior.

BIS Soil Classification

Soil is identified and classified in an appropriate group on the basis of grading and plasticity after excluding boulders and cabbies. Each group is represented by a group symbol having primary and secondary descriptive letters.

Soil component	Primary letters
Gravel	G
Sand	S
Silt	M
Clay	C
Organic matter	O
Highly organic soil (peat)	P_t

Secondary letters	Description
W	Well graded, clean
P	Poorly graded, Fairly clean
L	Low plasticity (w_L <50 %)
I	Medium plasticity (W_L between 35-50 %)
H	High plasticity (w_L > 50%)

Soils are broadly divided into three divisions by BIS:

Coarse Grained Soils

Soils in which more than half the total material by weight is larger than 75 micron IS sieve, are called coarse grained soils.

Coarse grained soil is further divided into two subdivisions:

Gravels

Soils in which more than half the coarse fraction (+75 micron) is larger than 4.75 mm, are called gravels (G).

Gravely soils have following four groups: Symbol

(a) Well graded gravels with little or no fines – GW

(b) Poorly graded gravels with little or no fines – GP

(c) Silty gravels – GM

(d) Clayey gravels – GC

Sands

Soils in which more than half the coarse fraction (+75 micron) is smaller than 4.75 mm, are called sands (s).

Sandy soils have following four groups:

(a) Well graded sands with little or no fines – SW

(b) Poorly graded sands with little or no fines – SP

(c) Silty sands – SM

(d) Silty clay – SC

Fine Grained Soils

Soils in which more than half of the total material by weight is smaller than 75 micron IS sieve, are called fine grained soils.

Fine grained soils are further divided into three subdivisions on the basis of liquid limit:

Silts and Clays of Low Compressibility (L)

Having liquid limit less than 35%.

Silts and Clays of Medium Compressibility (I)

Having liquid limit lie between 35 to 50%.

Silts and Clays of High Compressibility (H)

Having liquid limit greater than 50%.

Groups

The coarse grained soils are further divided into eight basic soil groups and the fine grained soils are divided into nine basic soils groups.

Fine Grained Soils have following Groups:

Low Compressibility Fine Grained Soils

(a) Inorganic silts of low compressibility – ML

(b) Inorganic clays of low compressibility – CL

(c) Organic soil (silts and clays) of low compressibility – OL

Medium Compressibility Fine Grained Soils

(a) Inorganic silts of medium compressibility – ML

(b) Inorganic clays of medium compressibility – CI

(c) Organic soil of medium compressibility – OI

High Compressibility Fine Grained Soils

(a) Inorganic silts of high compressibility – MH

(b) Inorganic clays of high compressibility – CH

(C) Organic soil of high compressibility – OH

Highly Organic Soils and other Miscellaneous Soils Materials

These soils have large percentage of fibrous organic matter, such as peat and particles of decomposed vegetation. In addition, certain soils containing shells, cinders and other non-soil materials in sufficient quantities are also grouped in this division.

Coarse grained and fine grained soils are further divided into subdivisions as given below:

Plasticity Chart

Plasticity chart is used to classify fine grained soils figure below hows a plasticity chart.

A line on the plasticity chart has the following linear equations: IP

$$= 0.73 \ (W_L - 20)$$

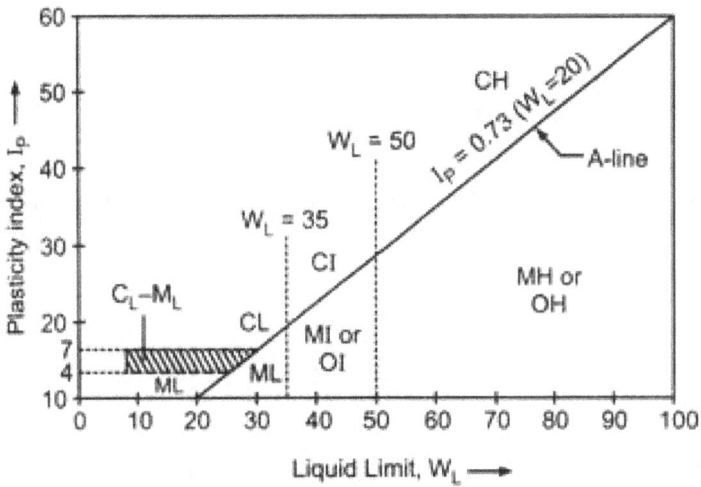

Figure: Plasticity chart

Inorganic clays lie above the A-line. Inorganic silts and organic soils lie below the A-line. Soils which plot above the A-line and which have plasticity index between 4 and 7, represents a boarder-line case and represented by dual symbol, ML – CL.

Soil Porosity

Pore-spaces (also called voids) in a soil consist of that portion of the soil volume not occupied by solids, either mineral or organic. The pore-space under field conditions, are occupied at all times by air and water. Pore-spaces directly control the amount of water and air in the soil and indirectly influence the plant growth and crop production.

Types of Pores in Soil

In general there are broadly two types of pores in soils:

(i) Macro-pores and

(ii) Micro or capillary pores.

Macro Pores

Large-sized pores are referred to as macro-pores which allow air and water movement easily. Sands and sandy soils have a large number of macro-pores. It is found in between the granules.

Micro or Capillary Pores

Smaller sized pores are generally referred to as a micro or capillary pores in which movement of air and water is restricted to some extent. Clays and clayey soils have a greater number of micro or capillary pores. It has got more important in the plant growth relationship. It is found within the granules.

Besides soil pores have been divided into following four categories based on the size groupings of soil separates

Coarse Pores

Greater than 0.2 mm or 200 microns (0.008 inch) average diameter.1 micron = 1 millionth of a metre.

Size of Medium Sands

Medium Pores

0.2-0.02 mm or 200-20 microns (0.008-0.0008 inch) average diameter. Size of coarse silt particles.

Fine Pores

0.02-0.002 mm, which is 20-2 microns (0.0008 inch) average diameter? Size of fine silt particles.

Very Fine Pores

Less than 2 microns (0.00008 inch) average diameter (Large clay particles are 2 microns in average diameter). Size of large clay particles. Porosity refers to the percentage of soil volume occupied by pore spaces. Size of individual pores, rather than total pore space in a soil, is more significant in its plant growth relationship.

For optimum growth of the plant, the existence of approximately equal amount of macro and micro-pores which influence aeration, permeability, drainage and water retention favorably. Porosity of a soil can be easily changed.

Factors Affecting Porosity of Soil

Wide difference in the total pore space of various soils occurs depending upon the following several factors:

- Soil Structure

A soil having granular and crumb structure contains more pore spaces than that of prismatic and platy soil structure. So well aggregated soil structure has greater pore space as compared to structure less or single grain soil.

- Soil Texture

In sandy soils the total pore space is small whereas in fine textured clay and clayey loam soils total pore space is high and there is a possibility of more granulation in clay soils.

- Arrangement of Soil Particles

When the sphere like particles is arrangement in columnar form (i.e. one after another on the surface forming column like shape) it gives the most open packing system resulting very low amount of pore spaces. When such particles are arranged in the pyramidal form it gives the most close packing system resulting high amount of pore spaces.

- Organic Matter

Soil containing high organic matter possesses high porosity because of well aggregate formation.

- Macro-Organisms

Macro-organisms like earthworm, rodents, and insects etc. increase macro-pores in the soil.

- Depth of Soil

With the increase in depth of soil, the porosity will decrease because of compactness in the sub-soil.

- Cropping

Intensive crop cultivation tends to lower the porosity of soil as compared to fallow soils. The decrease in porosity may be due to reduction in organic matter content.

- Puddling

Due to puddling under sufficient soil moisture, the soil surface layer is made dense and compact. Eventually, the porosity of this surface soil is reduced by the infiltration of muddy surface materials.

Atterberg Limits

A fine-gained soil can exist in any of several states; which state depends on the amount of water in the soil system. When water is added to a dry soil, each particle is covered with a film of adsorbed water. If the addition of water is continued, the thickness of the water film on a particle increases. Increasing the thickness of the water films permits the particles to slide past one another more easily. The behavior of the soil, therefore, is related to the amount of water in the system. Approximately sixty years ago, A. Atterberg defined the boundaries of four states in terms of "limits" as follows:

- Liquid limit: The boundary between the liquid and plastic states;

- Plastic limit: The boundary between the plastic and semi-solid states;

- Shrinkage limit: The boundary between the semi-solid and solid states.

These limits have since been more definitely defined by A. Casagrande as the water contents which exist under the following conditions:

- Liquid limit: The water content at which the soil has such a small shear strength that it flows to close a groove of standard width when jarred in a specified manner.

- Plastic limit: The water content at which the soil begins to crumble when rolled into threads of specified size.

- Shrinkage limit: The shrinkage limit (SL) is the water content where further loss of moisture will not result in any more volume reduction.

The Liquid Limit, also known as the upper plastic limit, is the water content at which soil changes from the liquid state to a plastic state.

It is the minimum moisture content at which a soil flows upon application of very small shear force. OR

The moisture content at which any increase in the moisture content will cause a plastic soil to behave as a liquid. The limit is defined as the moisture content, in percent, required to close a distance of 0.5 inches along the bottom of a groove after 25 blows in a liquid limit device.

Liquid Limit (LL or wL) - the water content, in percent, of a soil at the arbitrarily defined boundary between the semi-liquid and plastic states.

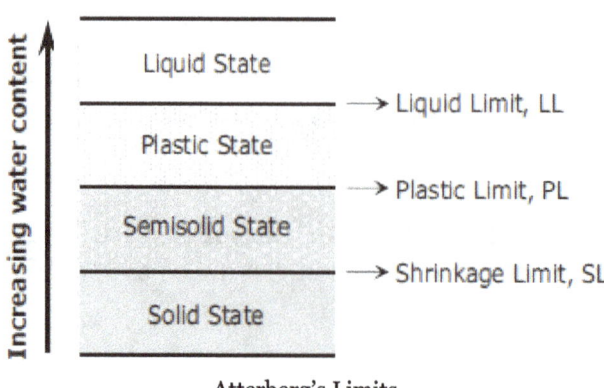

Atterberg's Limits

Plastic Limit

The Plastic Limit, also known as the lower plastic limit, is the water content at which a soil changes from the plastic state to a semisolid state.

The moisture content at which any increase in the moisture content will cause a semi-solid soil to become plastic. The limit is defined as the moisture content at which a thread of soil just crumbles when it is carefully rolled out to a diameter of 1/8 inch.

Plastic Limit (PL or wP) - the water content, in percent, of a soil at the boundary between the plastic and semi-solid states.

Plasticity

It is defined as the property of cohesive soil which posses the ability to undergo changes of shape without rupture or a change in volume.

Non Plastic Soils

Soil that do not have plastic limit are reported as being non-plastic e.g fairly clean sand, rock dust etc. Non plastic soils make excellent road materials when properly confined under wearing course .e.g. Rock dust. Even when wet they form hard, durable surface whereas clean sand displaces easily under the load. However, their use for base course or for fill brings difficult construction problems.

Plasticity Index (PI)

Plasticity Index is the range of water content over which a soil behaves plastically. It indicates the degree of plasticity of the soil. PI is the difference between the liquid limit and the plastic limit. Greater the difference, greater is the plasticity of the soil. Soils with a high PI tend to be predominantly clay, while those with a lower PI tend to be predominantly silt. Soils with high plasticity index are highly compressible. Plasticity index in also a measure of cohesiveness with high value of PI indicating high degree of cohesion. Experience shows that soils with high P.I are much less desirable for sub grade or base course than those having less indexes. For base course common specification is P.I < 3.

Shrinkage Limit

It is the maximum water content at which a reduction in water content will not cause a decrease in volume of the soil mass. It is the lowest water content at which soil can still be saturated.

Liquidity Index / Water Plasticity Ratio

The Atterberg's limits are found for remoulded soil samples. These limits as such do not indicate the consistency of undisturbed soil. The index to indicate consistency of undisturbed soils is called liquidity index of the soil or Water Plasticity Ration of the soil. Its advantage is that. The liquidity index (LI) is used for scaling the natural water content of a soil sample to the limits.

Flow Index

The curve obtained from the graph of water content against the log of blows while determining the liquid limit lies almost on a straight line and is known as the flow curve. The equation for flow curve is:

Where 'If is the slope of flow curve and is termed as "Flow Index".

Toughness Index

The shearing strength of clay at plastic limit is a measure of its toughness. It is the ratio of plasticity index to the flow index. It gives us an idea of the shear strength of soil.

Determining Liquid Limit and Plastic Limit of Soil

Apparatus

Liquid Limit Device: a mechanical device consisting of a brass cup suspended from a carriage designed to control its drop onto a hard rubber base. The device may be operated by either a hand crank or electric motor.

Cup: brass with mass (including cup hanger) of 185 to 215 g.

Cam: designed to raise the cup smoothly and continuously to its maximum height, over a distance of at least 1800 of cam rotation, without developing an upward or downward velocity of the cup when the cam follower leaves the cam.

Flat Grooving Tool: a tool made of plastic or non-corroding metal having specified dimensions.

Gage: A metal gage block for adjusting the height of the drop of the cup to 10 mm.

Ground Glass Plate: used for rolling plastic limit threads.

Significance and use

This testing method is used as an integral part of several engineering classifications systems to characterize the fine-grained fractions of soils and to specify the fine-grained fraction of construction materials. The liquid limit, plastic limit and plasticity index of soils are also used extensively, either individually or together, with other soil properties to correlate with engineering behavior such as compressibility, permeability, compactibility, shrink-swell and shear strength.

Procedure for Liquid Limit Test

1. Place a portion of the prepared sample in the cup of the liquid limit device at the point where the cup rests on the base and spread it so that it is 10mm deep at its deepest point. Form a horizontal surface over the soil. Take care to eliminate air bubbles from the soil specimen. Keep the unused portion of the specimen in the storage container.

2. Form a groove in the soil by drawing the grooving tool, beveled edge forward, through the soil from the top of the cup to the bottom of the cup. When forming the groove, hold the tip

of the grooving tool against the surface of the cup and keep the tool perpendicular to the surface of the cup.

3. Lift and drop the cup at a rate of 2 drops per second. Continue cranking until the two halves of the soil specimen meet each other at the bottom of the groove. The two halves must meet along a distance of 13mm (1/2 in).

4. Record the number of drops required to close the groove.

5. Remove a slice of soil and determine its water content, w.

6. Repeat steps 1 through 5 with a sample of soil at a slightly higher or lower water content. Whether water should be added or removed depends on the number of blows required to close the grove in the previous sample.

The liquid limit is the water content at which it will takes 25 blows to close the groove over a distance of 13 mm. Run at least five tests increasing the water content each time. As the water content increases it will take less blows to close the groove.

Sample number	01	02	03
Container number	24	21	25
Number of Blows	17	25	34
Mass of empty container (M1), gm	44.9	46	44.6
Mass of container + wet soil (M2), gm	78.3	81.3	76.8
Mass of container + dry soil (m³), gm	70	75.30	74.10
Water content= w = (M2 - m³/ m³ - M1) x 100, %	33.07	20.30	10.0

Liquid Limit, LL

Plot the relationship between the water content, w, and the corresponding number of drops, N, of the cup on a semi-logarithmic graph with water content as the ordinates and arithmetical scale, and the number of drops on the abscissas on a logarithmic scale. Draw the best fit straight line through the five or more plotted points. Take the water content corresponding to the intersection of the line with the 25 drop abscissa as the liquid limit, LL, of the soil.

Precautions

- After performing each test the cup and grooving tool must be cleaned.

- The number of blows should be just enough to close the groove.
- The number of blows should be between 10 and 40.

Applications

- The value of liquid limit helps in classification of fine grain soil.
- The values of liquid limit are required to calculate flow index, toughness index etc.

Procedure for Determination of the Plastic Limit

1. From the 20g sample select a 1.5 to 2 g specimen for testing.
2. Roll the test specimen between the palm or fingers on the ground glass plate to from a thread of uniform diameter.
3. Continue rolling the thread until it reaches a uniform diameter of 3.2mm or 1/8 in.
4. When the thread becomes a diameter of 1/8 in. reform it into a ball.
5. Knead the soil for a few minutes to reduce its water content slightly.
6. Repeat steps 2 to 5 until the thread crumbles when it reaches a uniform diameter of 1/8 in.
7. When the soil reaches the point where it will crumble, and when the thread is a uniform diameter of 1/8", it is at its plastic limit. Determine the water content of the soil.

Repeat this procedure three times to compute an average plastic limit for the sample.

Calculations

Plastic Limit, PL

Compute the average of the water contents obtained from the three plastic limit tests. The plastic limit, PL, is the average of the three water contents.

Plasticity Index

Calculate the plasticity index as follows: PI = LL - PL where:

LL = liquid limit, and PL = plastic limit.

Precautions

- The apparatus required for the experiment should be clean.
- All the readings should be noted carefully.
- Practical applications
- The value of liquid limit and plastic limit are used to classify fine grained soil.

- The values of liquid limit and plastic limits are used to calculate flow index, toughness index and plasticity index of the soil.

Observations and Calculations

Number of container	24	25
Weight of empty container(M1) gm	45.7	44.5
Weight of container + wet soil (M2) gm	50.6	49.5
Weight of container + dry soil (m³) gm	48	46
Water content = (M2- m³/ m³-M1)100 %	1.13	1.

Average plastic limit = (1.13 + 1.7)/2 =1.15%

Fall Cone Test

The fall cone test (FCT) is a simple testing method in which a cone is penetrated into a soil specimen by its self-weight and the penetration depth is measured. This test is extensively used for measuring Atterberg limits, i.e., liquid limit (w_L) and plastic limit (w_P), undrained shear strength (s_u) for intact as well as remolded clay sample and to determine sensitivity (S_t). On the other hand, Japan has been very conservative about the use of the FCT, and decided in 1997 to use it only to measure the liquid limit (w_L). However, w_L is in most cases still measured by the Casagrande method, i.e., using a cup.

Japan is one of only a few countries where the su design value is determined only by the unconfined compression test (UCT). The theoretical background for the validity of a half of the unconfined compression strength ($q_u/2$) as the mobilized undrained shear strength (s_{um}) has been studied by many researchers. According to their studies, $q_u/2$ is a suitable value that balances the under- and over-evaluating factors well to indicate true strength, such as sample disturbance, anisotropy and the rate effects. In other countries, however, the UCT is considered a less reliable test for obtaining sum. Indeed, the Eurocode 7 describes the UCT as an index test with the same rank of the FCT.

S_t is a very important parameter in geotechnical engineering, especially in practical construction works, i.e., to evaluate traffic ability or workability. S_t used to be measured in Japan by the UCT, in which the qu value is measured for intact and remold conditions and S_t was defined as a ratio of these q_u values. However, this article was eliminated in the standard of JGS in 1976, because it is very difficult to make a remolded specimen with highly sensitive samples because of their low strength, and for a very stiff sample, thoroughly remolding is very difficult. A testing method still in use for determining S_t is the Field Vane Test (FVT). In this standard, S_t is defined as a strength ratio of undisturbed and disturbed strengths after more than 10 turns of the vane blade.

Testing Methods

Fall Cone Test (FCT)

The FCT was carried out following the standard of JGS for the measurement of w_L, i.e., the mass and the conical angle of the cone are 60 g and 60°, respectively. These values are the same as those in Canada and Scandinavian countries. After the tip of the cone touches the specimen surface, the cone was freely dropped. After 5 s, the penetration depth was measured by a dial gage. The su was calculated by the following equation:

$$S_u = K_\alpha \left(mg / d^2 \right)$$

where m is mass of the cone (=60 g), g is the earth gravity acceleration, d is the penetration depth, and kα is the cone factor depending on the cone angle. k_α=0.29 was adopted, according to Wood for the cone angle of 60° in this study.

Intact specimens for FCT were obtained by a soil sample extruded from a sampling tube and cut by a wire saw (thickness=approx. 40 mm). A cylindrical specimen was directly placed on the basement of the Fall Cone apparatus, and the penetration test was carried out at 5 different points of the same specimen, where each point was far enough from the previous penetration tests not to be influenced by disturbance. The represented value from FCT was adopted as the average values of three measured ones, omitting the maximum and minimum values. The remolded specimens were made as follows: soil fragments yielded from the trimming of specimens for other mechanical tests or soil samples after testing of UCT were put into a plastic bag and kneaded uniformly by hands for 5 min. The kneaded sample was then stuffed into a cup defined by the standard of liquid limit by the fall cone (JGS 0142-2009), i.e., its inside diameter and depth are more than 60 mm and 30 mm, respectively. The remolded strength (S_{uFCT}) was calculated from one penetration depth, because the measured value was not scattered unlike intact samples.

Unconfined Compression Test (UCT)

The UCT was also carried out, following the standard by the Japanese Industrial Standard (JIS A 1216:2009): the specimen was trimmed by a wire saw to a diameter and height of 35 and 80 mm, respectively. The specimen was compressed at a constant strain rate of 1%/min. The peak strength was defined as the unconfined compression strength (qu).

Field Vane Test (FVT)

The testing method for the FVT mainly followed the JGS 1411-2003, using a penetration type without a borehole. The vane blade was 40 mm in diameter and 80 mm in height, respectively. To avoid damage the vane blade was protected by the sheath in the process of the penetration. When the device arrived at a testing depth, only the vane blade penetrated from the sheath into the ground. Then the blade was rotated at a rotation speed of 6°/min. After attaining the peak strength (s_{uFVT}), the vane was rotated a further 30 turns at a rapid rotation speed (about 5 s per a turn). Again the vane blade was rotated at the same rotation speed of 6°/min to obtain the remolded strength (s_{uFVT30}).

Laboratory Vane Test (LVT)

The LVT was also performed, using a smaller vane (the diameter and height were 20 mm and 40 mm, respectively). For measuring undisturbed strength (s_{uLVT}), the vane was inserted into the soil sample kept in the sampling tube (the Japanese standard sampler, i.e., inside diameter=75 mm).

The remolded specimens were created by two methods. One is the same as the FVT, i.e., after attaining peak strength, the vane was rapidly rotated by 30 turns and the remolded strength (S_{uLVT30}) was measured by the rotation speed of 6°/min. Another remolded specimen was made by kneading soil samples in the same manner as the FCT. The remolded strength (s_{urLVT}) was measured by the LVT after the remolded sample was stuffed into a container.

Atterberg limits

Liquid and plastic limits (w_L and w_P, respectively) were obtained by the Japanese Industrial Standard, i.e., FCT was not used either for w_L or w_P.

Stresses in the Ground

Stresses in Dry Soil

Let A in figure below represent a small cubical shaped element of soil at a depth z in an extensive uniform soil deposit in which the ground surface is horizontal and which has been formed by the gradual accretion of material on the ground surface. Because the soil deposit is extensive (by comparison with distance z) in the horizontal direction the stresses on element A will be identical with the stresses on an adjacent element at the same depth below the ground surface. This means that there cannot be any shear stresses existing on the vertical or horizontal planes which bound element A. In other words the vertical stress (σ_v) and horizontal stress (σ_H) are principal stresses.

The vertical stress on element A can be determined simply from the mass of the overlying material. If ρd represents the density of the soil, the vertical stress is,

$$\sigma_v = \rho d \, gz$$

The horizontal stress is customarily expressed as a proportion of the vertical stress,

$$\sigma_H = K'_o \, \sigma_v = K'_o \, \rho d \, gz$$

where K'_o = coefficient of earth pressure at rest in terms of effective stresses.

This coefficient contains the words "at rest" since the soil was deposited under conditions of zero horizontal strain. In other words, because of the large lateral extent of the soil deposit, the vertical planes on any soil element A do not experience any lateral movement as the stresses increase as a consequence of the accretion of material on the ground surface. The distributions of σ_v and σ_H as a function of depth are illustrated in figure below.

The value of K'_o in equation $\sigma_H = K'_o \, \sigma_v = K'_o \, \rho d \, gz$ can be determined if the soil is assumed to behave as an elastic solid, as follows:

$$\varepsilon_H = \frac{1}{E}(\sigma_H - v\sigma_v - v\sigma_H) = 0$$

Where ε_H =horizontal strain

E,v= elastic parameters for the soil

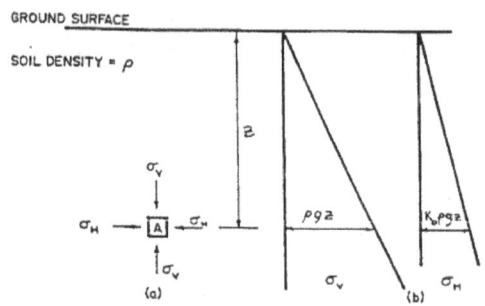

Figure: Stresses in Dry Soil

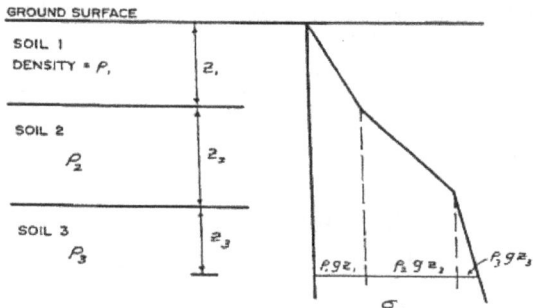

Figure: Stresses in a Layered Deposit

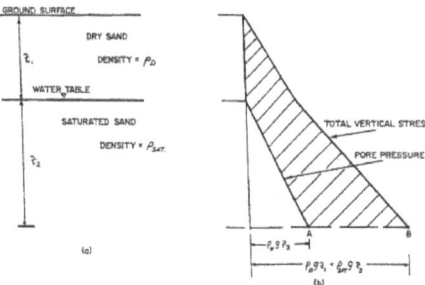

Figure: Stresses in a Saturated Soil

$$\therefore \sigma_H - v\sigma_v - v\sigma_H = 0$$

$$\sigma_H(1-v) = v\sigma_V$$

$$\therefore \frac{\sigma_H}{\sigma_v} = K'_o = \frac{v}{1-v}$$

which indicates that K'o varies from 0 to 1.0 as the Poisson's ratio varies from 0 to 0.5.

The assumption of elastic behaviour for many soils may be an unrealistic idealization so the value of K'o should be determined experimentally. In the laboratory, K'o can be determined by applying a vertical stress σ_v to a soil sample while preventing all horizontal movement. The value of the horizontal stress σ_H required to prevent this movement is measured. The value of K'o is then calculated from the measured value of σ_H and the applied value of σ_v.

The stresses in a deposit consisting of layers of soil having different densities may be determined by a simple extension of the technique described above.

vertical stress at depth $z_1 = \rho_1 g z_1$

vertical stress at depth $(z_1 + z_2) = \rho_1 g z_1 + \rho_2 g z_2$

vertical stress at depth $(z_1 + z_2 + z_3) = \rho_1 g z_1 + \rho_2 g z_2 + \rho_3 g z_3$

Principle of Effective Stress

Figure above represents a cross section through an extensive deposit of sand. The water table is present at a depth of z_1 below the ground surface. The sand below the water table is saturated and that above the water table is dry. It is assumed that the capillary rise in this soil is zero. The total vertical stress may be calculated by using the density above the water table and sat below the water table. This leads to an expression for the total vertical stress σ_v at a depth of $(z_1 + z_2)$ below the ground surface of,

$$\sigma_v = \rho d\, gz_1 + \rho_{sat}\, gz_2$$

Because of the presence of the water table a hydrostatic pressure will also be present in the pore water. This pore pressure (u) is zero at the water table and will increase linearly with increasing depth, the value at a depth of z_2 below the water table being,

$$u = \rho w g z_2$$

The values of the vertical stress, σ_V and pore pressure u (sometimes called neutral stress) at various depths are illustrated in the same diagram in figure above.

The difference between the total stress σ and the pore pressure (u) in a saturated soil has been defined by Terzaghi as the effective stress σ.

$$\sigma = \sigma - u$$

The implications of this definition are among the most important in soil mechanics. Changes in soil properties are governed by changes in the effective stress and not solely by changes in total stress or changes in pore pressure. Unlike the pore pressure, the effective stress and total stress are tensors. This means that direction as well as magnitude must be specified.

The effective vertical stress (σ_v) at a depth of $(z_1 + z_2)$ below the ground surface may be found from equations $\sigma_V = \rho d\, gz_1 + \rho_{sat}\, gz_2$ and $u = \rho w g z_2$, as follows,

$$\begin{aligned}
\sigma_V &= \sigma_V - u \\
&= \rho_d\, gz_1 + \rho_{sat}\, gz_2 - \rho_w\, gz_2 \\
&= \rho_d\, gz_1 + (\rho_{sat} - \rho_w)gz_2 \\
&= \rho_d\, gz_1 + \rho d\, gz_2
\end{aligned}$$

where $\rho_b = \rho_{sat} - \rho_w$ and is called the buoyant density or submerged density.

This value of the effective vertical stress, σ_v is represented by the line AB (stress at a depth of $(z_1 + z_2)$ below the ground surface) in figure above and the distribution of the effective vertical stress as a function of depth is shown by the shaded area in this figure. For depths of less than z1 below the ground surface the total and effective stresses are equal since the pore pressure is zero.

The effective horizontal stress σ_H in an extensive soil deposit may be found from equations

$$\sigma_H = K'_0\, \sigma_v = K'_0\, \rho d\, gz$$

and $\sigma = \sigma - u$ with one important modification. When pore pressures are present the stresses in equation $\sigma_H = K'_0\, \sigma_v = K'_0\, \rho d\, gz$ must be effective stresses. Note that the value of the coefficient of earth pressure at rest is defined in terms of effective stresses and not total stresses.

$$\sigma'_H = K'_o \sigma'_v$$
$$= K'_o(\sigma_v - u)$$

For the case illustrated in figure above the value of the horizontal effective stress at a depth of $(z_1 + z_2)$ below the ground surface is,

$$\sigma'_H = K'_o(\rho_d gz_1 + \rho_b gz_2)$$

Example

A reservoir of water 10m deep is underlain by an extensive sand deposit. Referring to figure above determine the change in effective vertical stress at point P if the water depth in the reservoir is increased to 18 m.

This problem will be solved in two ways:

(a) the initial effective vertical stress σ'_i at point P may be found from the initial total vertical, stress σ_i and the initial pore pressure ui,

$$\sigma'_i \quad = \sigma_i = u_i$$
$$\sigma_i \quad = 1000 \times 9.81 \times 10 + 2100 \times 9.81 \times 6 \ N/m^2$$
$$= 98.1 + 123.6 \ kN/m^2 \quad = 221.7 \ kN/m^2$$
$$u_i \quad = 1000 \times 9.81 \times 16 \ N/m^2 = 157.0 \ kN/m^2$$
$$\therefore \quad \sigma'_i \quad = 221.7 - 157.0 \qquad\qquad = 64.7 \ \ kN/m^2$$

Similarly the final effective vertical stress σ'_f may be found,

$$\sigma'_f \quad = \quad (1000 \times 9.81 \times 18 + 2100 \times 9.81 \times 6) - (1000 \times 9.81 \times 24)$$
$$= \quad 176.6 + 123.6 - 235.5 \ \ kN/m^2 = 64.7 \ \ kN/m^2$$

Since the initial and final values of effective vertical stress are the same, the change in effective vertical stress is zero.

The initial effective vertical stress may be calculated from the buoyant density ρ_b of the sand:

$$\rho_b \quad = \rho_{sat} - \rho_w$$
$$= 2100 - 1000 = 1100 \ \ kN/m^3$$

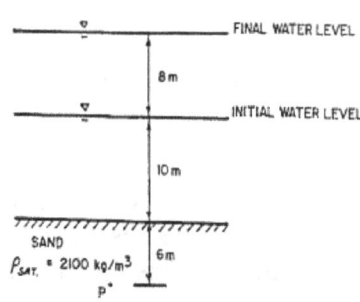

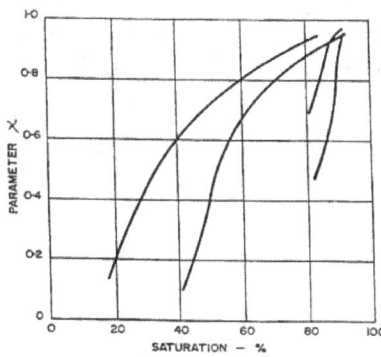

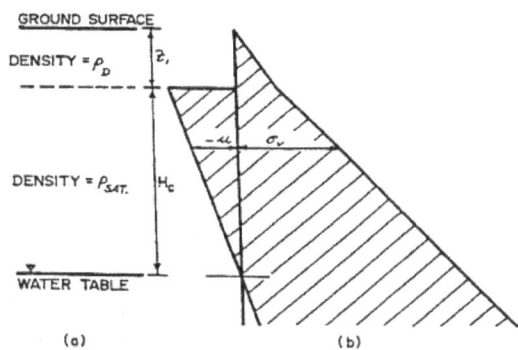

Figure: Measured χ Values for Four Clay Soils   Figure: Stresses in the Capillary Zone

$$\sigma_i' = 1100 \times 9.81 \times N/m^2$$
$$= 64.7\ kN/m^3$$

which is the same answer as in (a). When doing the calculation of effective stress in this way, it is clear that in the calculation of the final effective stress none of the terms in the equation has changed - the buoyant density remains the same and the depth of submerged sand of 6m does not alter. Consequently there can be no change in the effective stress as the reservoir depth increases.

In this problem the total stress and the pore pressure increased by exactly the same amount as the water depth increased. As long as the sand remains submerged this equality is maintained for any increase or decrease in water depth.

Effective Stress in Partly Saturated Soil

Equation $\sigma = \sigma - u$ cannot be used to determine effective stress in partly saturated soil because there are now two pore pressures instead of one - the pore water pressure (u_w) and the pore air pressure (u_a). it is clear that, in general, these two pore pressures will be unequal.

$$\sigma' = \sigma - u_a + \chi(u_a - u_w)$$

where χ is a parameter which may vary from 0 to 1.

When χ = 1 the soil is saturated and equation above simplifies to,

$$\sigma' = \sigma - u_w$$

When χ = 0 the soil is dry and equation $\sigma' = \sigma - u_a + \chi(u_a - u_w)$ simplifies to the same form as equation,

$$\sigma' = \sigma - u_a$$

The parameter χ has been found to depend mainly on the degree of saturation as illustrated in figure above but it has been suggested that it also depends on soil structure, the cycle of wetting and drying or stress change leading to the particular value of degree of saturation at which χ is measured. Some aspects of the χ concept remain to be explained since some measurements have yielded values outside of the expect range of 0 to 1.

Around 1960 several other effective stress equations for unsaturated soils were proposed, such as Aitchison and Jennings. Common to all equations was the incorporation of a soil parameter characteristic of the soil behavior. These soil parameters have been virtually impossible to evaluate uniquely and difficult to apply to practical problems)). As described by Fredlund and Morgenstern and more recently by Fredlund, there has been an increased tendency to uncouple the terms in the effective stress equation, $(\sigma - u_a)$ and $(u_a - u_w)$, and treat the stress variables independently.

Stresses in the Capillary Zone

Figure above represents a soil deposit in which the height of capillary rise H_c is less than the depth of the water table below the ground surface. To simplify the calculation it will be assumed that the soil is saturated within the capillary zone and completely dry above the height of the capillary rise.

The total vertical stress may be determined from an expression similar to equation $\sigma_V = \rho d\, g z_1 + \rho_{sat}\, g z_2$. At a depth of z_1, below the ground surface,

$$\sigma_V = \rho_d\, g z_1$$

and at the water table elevation

$$\sigma_V = \rho_d\, g z_1 + \rho_{sat}\, gH_c$$

The distribution of σ_v as a function of depth,

The pore pressure is zero down to a depth of z_1 below the ground surface. At this depth there is a change in pore pressure in the capillary zone to a value of,

$$u = -\rho_w\, gH_c$$

and the pore pressure increases to zero at the water table elevation as shown in

Within the capillary zone the soil is saturated so the principle of effective stress as expressed in equation $\sigma' = \sigma - u$ may be applied

$$\sigma'_V = \sigma_V - u$$

At a depth of z_1, below the ground surface

$$\begin{aligned}\sigma_V &= \rho d\, g z_1 - (-\rho_w\, gH_c)\\ &= \rho d\, g z_1 + \rho_w\, gH_c\end{aligned}$$

and at the water table elevation

$$\sigma_i = \sigma_V = \rho d\, g z_1 + \rho_{sat}\, gH_c$$

The distribution of effective vertical stress is shown by the shaded area in

Non-perpendicular Stresses

In the cases of saturated soil the total stress and the pore pressure have been acting in the same direction so there has been no confusion over the use of the effective stress equation $\sigma' = \sigma - u$. It has already been stated that the total stress is a tensor so its direction must be taken into account. On the other hand, pore pressure is not a tensor and at any point it acts equally in all directions.

BC represents a plane in a mass of saturated soil. The total stress (σ) acts on this plane at an angle θ to the normal but the pore pressure (u) acts at right angles to the plane. In this case the effective stress cannot be determined simply by subtracting u from σ. It is necessary to consider the two components of the total stress σ, one perpendicular to the plane and the other parallel to it. The former component, the total normal stress (σ_n) is found by resolution in the normal direction.

$$\sigma_n = \sigma \cos \theta$$

Similarly the component parallel to the plane, the shear stress (τ) is,

$$\tau = \sigma \sin \theta$$

Now the effective normal stress $\left(\sigma_n' \right)$ can be found by subtracting the pore pressure from the total normal stress,

$$\sigma_n' = \sigma_n - u$$

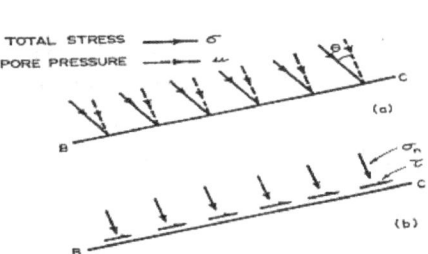

Figure: Stresses Acting on a Plane

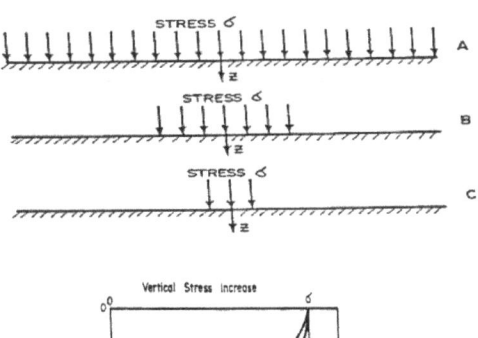

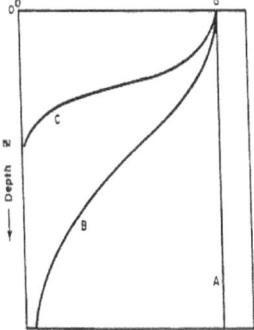

Figure: Effect of Lateral Extent of Surface Load on Stress Change in the Ground

and the shear stress is unaffected by the pore pressure. The components of the total stress acting on plane BC are,

(a) an effective stress $\left(\sigma'_n\right)$ acting normal to the plane,

(b) a pore pressure (u) acting normal to the plane, and

(c) a shear stress (τ) acting parallel to the plane.

Transmission of Stress into the Ground

Here we will concentrate on stress changes produced by loads (forces and stresses) applied at the ground surface. In the case of uniform vertical stresses applied over a large area of the ground surface, the vertical stress changes at all depths in the ground are equal to the vertical stress applied at the ground surface as illustrated by case A in figure above In most soil engineering problems the load on the ground surface does not have a large lateral extent. In this case the stress change in the ground does not remain constant but decreases with increasing depth below the ground surface as shown in case B in figure above. In the case of a surface loading of very limited lateral extent a negligible stress change is experienced at great depths.

Quantitatively the stress changes at various depths are often calculated on the assumption that the solid behaves as a homogeneous, isotropic elastic solid. For such a solid, Boussinesq, in 1885 produced a solution for a concentrated load Q on the surface of the elastic solid. Using the symbols in figure above his solution for the vertical stress change at any point was.

$$\sigma_V = \frac{3Q\cos^5\theta}{2\pi z^2} = \frac{3Q}{2\pi} \cdot \frac{Z^3}{\left(r^2+z^2\right)^{5/2}} = \frac{Q}{z^2}l_1$$

Where,

$$l_1 = \frac{3}{2\pi}\left(\left(r/z\right)^2+1\right)^{-5/2}$$

This equation and table indicate the way in which the vertical stress change decreases with increasing distance from the concentrated load.

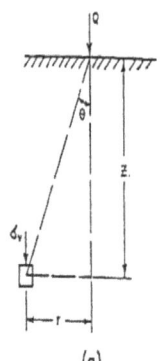
(a)

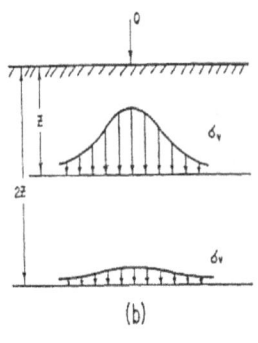
(b)

Stresses Produced by a Concentrated
Load at the Ground Surface

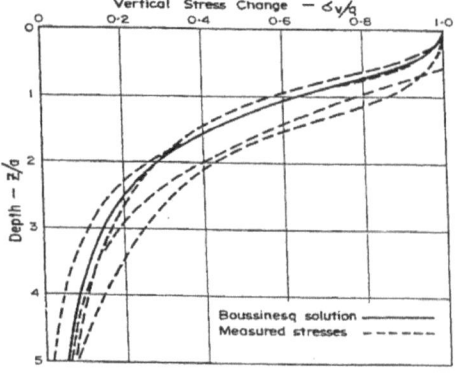

Measured and Calculated Stresses
for a Circular Loaded Area on Sand

(r/z)	I_1	(r/z)	I_1
0	0.4775	0.9	0.1083
0.1	0.4657	1.0	0.0844
0.2	0.4329	1.5	0.0251
0.3	0.3849	1.75	0.0144
0.4	0.3295	2.0	0.0085
0.5	0.2733	2.5	0.0034
0.6	0.2214	3.0	0.0015
0.7	0.1762	4.0	0.0004
0.8	0.1386	5.0	0.0001

For a pressure applied over a finite area on the surface a solution could be developed from equation $\sigma_v = \dfrac{3Q \cos^5 \theta}{2\pi z^2} = \dfrac{3Q}{2\pi} \cdot \dfrac{Z^3}{\left(r^2 + z^2\right)^{5/2}} = \dfrac{Q}{z^2} I_1$ by assuming that the area is subdivided into a number of small areas with a concentrated load acting at the center of each. For example, for a circular area of radius a and a uniform pressure of q at the surface, the expression for vertical stress change, σ_v at any depth z beneath the center of the circular area is

$$\sigma_v = q\left(1 - \frac{z^3}{\left(a^2 + z^2\right)^{3/2}}\right)$$

The way in which σ_v varies with (z/a) is given in table below.

(z/a)	$\left(\sigma v / q\right)$	(z/a)	$\left(\sigma v / q\right)$
0	1.0000	1.5	0.4240
0.05	0.9998	2.0	0.2845
0.10	0.9990	2.5	0.1996
0.2	0.9925	3.0	0.1436
0.5	0.9106	4.0	0.0869
1.0	0.6465	5.0	0.0571

Measured and calculated stress for a circular loaded area on sand.

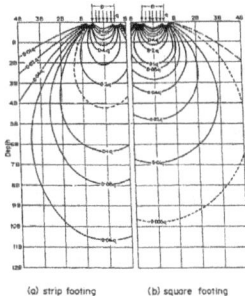

(a) strip footing   (b) square footing

Figure: Contours of Equal Vertical Stress

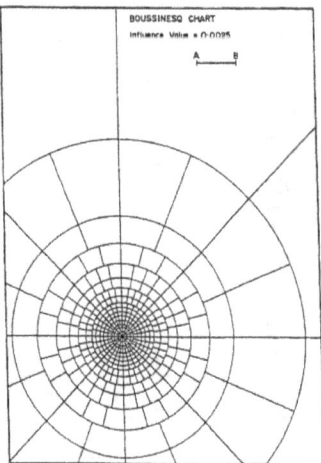

Figure: Chart for Determination of Vertical Stress

Unfortunately, the level of agreement for measured and calculated radial and tangential stresses is much poorer than that indicated in figure above for vertical stresses.

Elastic solutions are also available for loaded areas of different shapes. Figure contours of equal vertical stresses illustrates the patterns of vertical stress (isobars) produced by a uniformly loaded strip footing (infinitely long loaded area of width B) and a uniformly loaded square footing. This figure shows that the vertical stress change penetrates deeper into the ground in the case of a strip footing compared with the case of a square footing.

For loaded areas of irregular shape the simplest means of calculating the vertical stress change at any

depth is by means of a numerical integration procedure for equation $\sigma_V = \dfrac{3Q}{2\pi z^2} \cos^5 \theta = \dfrac{3Q}{2\pi} \cdot \dfrac{z^3}{\left(r^2 + z^2\right)^{5/2}} = \dfrac{Q}{z^2} I_1$

by means of influence charts of the type proposed by Newmark. In figure above shows chart for determination of vertical stress. The stress is determined by carrying out the following steps:

(a) Sketch the outline of the loaded area to scale such that the length AB is equal to the depth at which the stress is to be calculated,

(b) The sketch is placed on the chart with the center of the concentric circles coincident with the horizontal location for which the stress is to be calculated,

(c) Count the number of chart subdivisions (influence areas) enclosed within the sketch of the loaded area,

(d) calculate the stress change as follows:

stress change σ_V = number of influence areas

x influence value for the chart

x pressure acting on loaded area.

In addition to what has been mentioned above, there is now a wide variety of solutions available for stresses in non-homogeneous and anisotropic materials.

A simple but approximate method is sometimes used for calculating the stress change at various depths as a result of the application of a pressure at the ground surface. This method is illustrated in figure above which a surface pressure (σ_v) acts on a strip footing of width B. The vertical stress change at any depth (z) is calculated by assuming that the surface force is uniformly distributed over an imaginary footing having a width of (B + z). That is, the transmission of stress is assumed to follow outward fanning lines at a slope of 1 horizontal to 2 vertical.

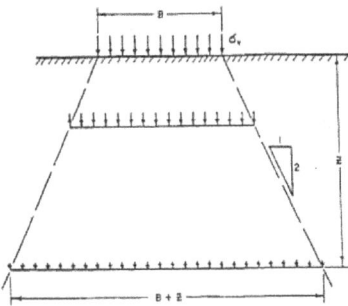

Figure: Stress Transmission by Means of the 2:1 Distribution

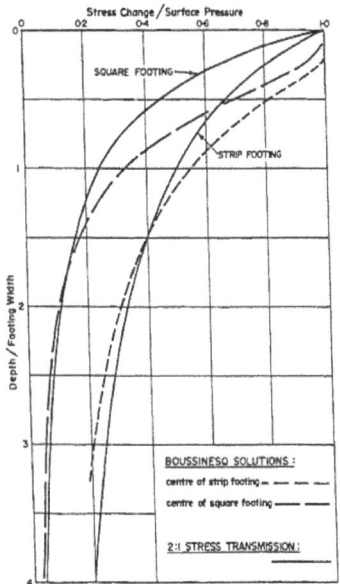

Figure: Stress Changes due to a Surface Pressure

For a strip footing of width B the vertical stress change ($\Delta\sigma_z$) at any depth z is,

$$\Delta\sigma_z = \frac{\sigma_v B}{(B + z)} = \frac{\sigma_v}{(1 + z/B)}$$

For a square footing of width B the vertical stress change ($\Delta\sigma_z$) at any depth z is,

$$\Delta\sigma_z = \frac{\sigma_v B^2}{(B + z)^2} = \frac{\sigma_v}{(1 + z/B)^2}$$

The stress changes according to the 2:1 transmission in equations above are compared with the elastic solutions in figure above. This figure shows that the 2:1 stress transmission results in an under-estimate of the stress change beneath the center of the loaded area at shallow depths. However the level of agreement is considerably improved if the average stress change beneath the whole of the loaded area is considered.

Example

Using the 2:1 stress distribution calculates the vertical stress change at a depth of 5 m for the following cases:

(a) a stress of 100kPa on a circular surface footing 2 m in diameter.

(b) a load of 10kN on a surface rectangular footing 2 m x 3 m.

(c) a stress of 200kPa on a surface strip footing 1m wide.

Surface load $= 100 \times \pi \times 2^2 / 4$

stress change at 5 m depth

$$\Delta\sigma_v \frac{\text{Surface load}}{\pi \times 7^2/4} = 8.2 \text{ kPa}$$

Stress change

$$\Delta\sigma_v = \frac{10}{(2+5)(3+5)}$$
$$= 0.18 \text{ kPa}$$

Stress change

$$\Delta\sigma_v = 200 \times 1/(1+5)$$
$$= 33.3 \text{ kPa}$$

Lateral Earth Pressures

Lateral Earth Pressure at Rest

Consider a vertical wall of height H, as shown in figure below, retaining a soil having a unit weight of γ. A uniformly distributed load, q/unit area, is also applied at the ground surface. The shear strength, s, of the soil is,

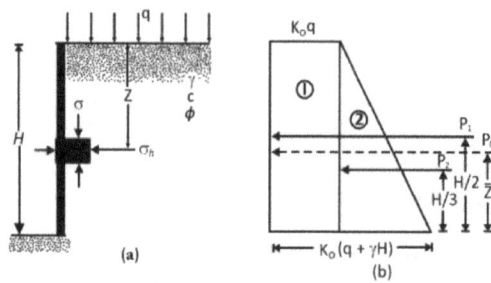

Figure: At-rest earth pressure

$$s = +\sigma' \tan \phi$$

Where,

c = cohesion

ϕ = angle of friction

σ' = effective normal stress

At any depth z below the ground surface, the vertical subsurface stress is

$$\sigma_v = q + yz$$

If the wall is at rest and is not allowed to move at all either away from the soil mass or into the soil mass (e.g., zero horizontal strain), the lateral pressure at a depth z is

$$\sigma_h = K_0 \sigma_v' + u$$

Where,

u = pore water pressure

K_0 = coefficient of at – rest earth pressure

For normally consolidated soil, the relation for K'_0 is,

$$K_0 \approx 1 - \sin \phi$$

Where,

ϕ = drained peak friction angle

Based on Brooker and Ireland's experimental results, the value of K_0 for normally consolidated clays may be approximated correlated with the plasticity index(Pl):

$$K_0 = 0.4 + 0.007(Pl) \qquad (\text{For Pl between 0 and 40})$$

And

$$K_0 = 0.64 + 0.001(Pl) \qquad (\text{For Pl between 40 and 80})$$

For over consolidated clays,

$$K_{0(overconsolidated\ clays,)} \approx K_{0(normally\ consolidated)} \sqrt{OCR}$$

Where

OCR = over consolidation ratio

Mayne and Kulhawy analyzed the results of 171 different laboratory tested soils. Based on this study, they proposed a general empirical relationship to estimate the magnitude of K_0 for sand and clay:

$$K_0 = (1 - \sin \phi) \left[\frac{OCR}{OCR_{max}^{(1-\sin \phi)}} + \frac{3}{4} \left(1 - \frac{OCR}{OCR_{max}} \right) \right]$$

Where,

OCR = present over consolidation ratio

OCR_{max} = maximum over consolidation ratio

In figure below OCR_{max} max is the value of OCR at point B.

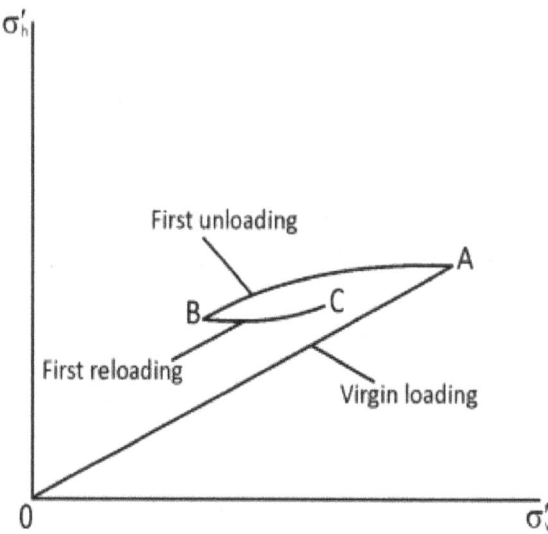

Figure: Stress history for soil under $KKoo$ condition

With a properly selected value of the at-rest earth pressure coefficient, equation can be used to determine the variation of lateral earth pressure with depth z. the variation of σ_h with depth for the wall. Note that if the surcharge $q = 0$ and the pore water pressure $u = 0$, the pressure diagram will be a triangle. The total force P_o, per unit length of the wall can now be obtained from the area of the pressure diagram as,

$$P_o = p_1 + p_2 = qK_oH + \frac{1}{2}\gamma H^2 K_o$$

Where,

p_1 =area of rectangle 1

p_2 = area of rectangle 1

The location of the line of action of the resultant force, P_o, can be obtained by taking the moment about the bottom of the wall. Thus

$$\bar{z} = \frac{p_1\left(\dfrac{H}{2}\right) + p_2\left(\dfrac{H}{3}\right)}{P_o}$$

If the water table is located at depth $z < H$ the at-rest pressure diagram shown in figure below will have to be somewhat modified. If the effective unit weight of soil below the water table equal $\gamma'\left(\text{that is, } \gamma_{sat} - \gamma_w\right)$

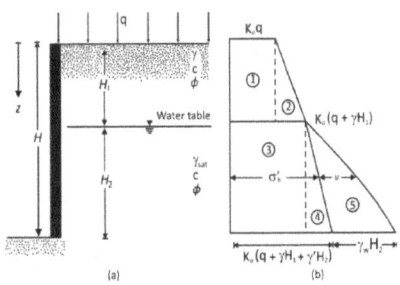

(a)   (b)

$$At\ z = 0, \sigma'_h = K_o\ \sigma'_v = K_o\ q$$

$$At\ z = H_1, \sigma'_h = K_o\ \sigma'_v = K_o\left(q + \gamma H_1\right)$$

$$At\ z = H_2, \sigma'_h = K_o\ \sigma'_v = K_o\left(q + \gamma H_1 + \gamma' H_2\right)$$

Note that in the preceding equations σ'_v and σ_h are effective vertical and horizontal pressures. Determining the total pressure distribution on the wall requires adding the hydrostatic pressure. The hydrostatic pressure, u is zero form z=o and $H_{1,}$ at z H_2 , $u = H_2 y_w$. the variation of σ_h and u with depth is shown in Hence the total force per unit length of the wall can be determined from the area of the pressure diagram. Thus,

$$P_o = A_1 + A_2 + A_3 + A_4 + A_5$$

Where,

A=area of the pressure diagram,

So,

$$P_o = K_o\ qH_1 + \frac{1}{2}K_o\ \gamma H_1^2 + K_o\left(q + \gamma H_1\right)H_2 + \frac{1}{2}K_o\ \gamma' H_2^2 + \frac{1}{2}\gamma_w H_2^2$$

Sheriff showed by several laboratory model tests that equation gives good results for estimating the lateral earth pressure at rest for loose sands. However, for compacted dense sand, it grossly underestimates the value of K_o. For that reason, they proposed a modified relationship for K_o :

$$K_o = \left(1 - \sin\phi\right) + \left(\frac{\gamma_d}{\gamma_d(\min)} - 5\right)5.5$$

Where,

γ_d = in situ unit weight of sand

$\gamma_{d(\min)}$ = minimum possible dry unit weight of sand

For the retaining wall shown in figure below, determine the lateral earth fore at rest per unit length of the wall. Also determine the location of the resultant force.

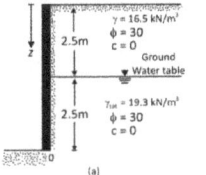
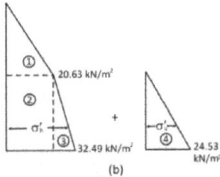

(a)   (b)

Solution

$$K_O = 1 - \sin\phi = 1 - \sin 30^\circ = 0.5$$
$$\text{At } z = 0, \sigma'_V = 0; \sigma'_h = 0$$
$$\text{At } z = 2.5m, \sigma'_V = (16.5)(2.5) = 41.63\,KN/m^2;$$
$$\sigma'_h = K_O \sigma'_V = (0.5)(41.25) = 20.63\,KN/m^2$$
$$\text{At } z = 5m, \sigma'_V = (16.5)(2.5) + (19.3 - 9.81)2.5 = 64.98\,KN/m^2;$$
$$\sigma'_h = K_O \sigma'_V = (0.5)(64.98) = 32.49\,KN/m^2$$

The hydrostatic pressure distribution is as follows:

From $z = 0$ to $z = 2.5m, u = 0$. At $z = 5m, u = \gamma_w(2.5) = (9.81)(2.5) = 24.53\,KN/m^2$ The pressure distribution for the wall.

The total force per unit length of the wall can be determined from the area of the pressure diagram, or

$$P_o \text{Area } 1 + \text{Area } 2 + \text{Area } 3 + \text{Area } 4$$

$$= \frac{1}{2}(2.5)(20.63) + (2.5)(20.63) + \frac{1}{2}(2.5)(32.49 - 20.63) + \frac{1}{2}(2.5)(24.53) = 122.85\,KN/m$$

The location of the center of pressure measured from the bottom of the wall (Point o) =

$$\bar{z} = \frac{(Area1)\left(2.5 + \dfrac{2.5}{3}\right) + (Area2)\left(\dfrac{2.5}{2}\right)(Area3 + Area4)\left(\dfrac{2.5}{3}\right)}{P_o}$$

$$= \frac{(25.788)(3.33) + (51.575)(1.25)(14.825 + 30.663)(0.833)}{122.85}$$

$$= \frac{85.87 + 64.47 + 37.89}{122.82} = 1.53m$$

Active Pressure

Ranking Active Earth Pressure

The lateral earth pressure conditions involve walls that do not yield at all. However, if a wall tends to move away from the soil a distance Δx, the soil pressure on the wall at any depth will decrease. For a wall that is frictionless, the horizontal stress σ_h at depth z will equal $K_o \sigma_v (K_o \gamma z)$ when Δx is zero. However, with $\Delta x > 0, \sigma_h$ will be less than $K_o \sigma_v$.

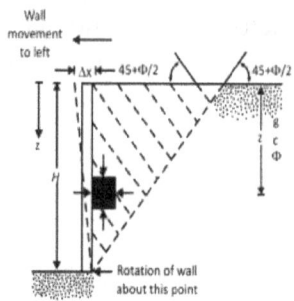

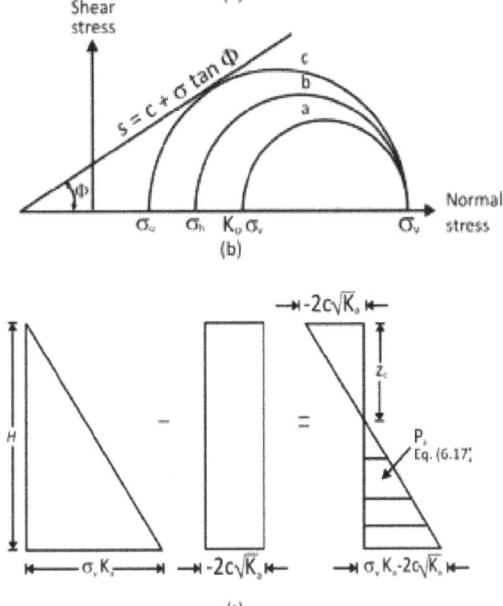

Figure: Rankine active pressure

The Mohr's circles corresponding to wall displacements of $\Delta x = 0$ and $\Delta x > 0$ are shown as circles a and b in figure (b). If the displacement of the wall, Δx continues to increase, the corresponding Mohr's circle eventually will just touch the Mohr-Coulomb failure envelope defined by the equation,

$$s = c + \sigma \tan \phi$$

This circle is marked c in figure (b). It represents the failure condition in the soil mass; the horizontal stress then equals σ_a. This horizontal stress, σ_a, is referred to as the Rankin active pressure. The slip lines (failure planes) in the soil mass will then make angles of $\pm (45 + \phi /2)$ with the horizontal.

The equation relating the principal stresses for a Mohr's circle that touches the Mohr-Coulomb failure envelope:

$$\sigma_1 = \sigma_3 \tan^2 \left(45 + \frac{\phi}{2} \right) + 2c \tan \left(45 + \frac{\phi}{2} \right)$$

For the Mohr's circle c in figure above,

Major Principal stress, $\sigma_1 = \sigma_v$

And

Minor Principal stress, $\sigma_3 = \sigma_a$

Thus,

$$\sigma_v = \sigma_a \tan^2 \left(45 + \frac{\phi}{2} \right) + 2c \tan \left(45 + \frac{\phi}{2} \right)$$

$$\sigma_a = \frac{\sigma_v}{\tan^2\left(45+\dfrac{\phi}{2}\right)} - \frac{2c}{\tan\left(45+\dfrac{\phi}{2}\right)}$$

Or

$$\sigma_a = \sigma_v \tan^2\left(45+\frac{\phi}{2}\right) - 2c \tan\left(45-\frac{\phi}{2}\right)$$

$$= \sigma_v K_a - 2c\sqrt{K_a}$$

Where,

$K_o = \tan^2\left(45 - \phi/2\right) =$ Rankine active pressure coefficient

The variation of the active pressure with depth for the wall shown in figure above is given in figure above. Note that $\sigma_v = 0$, at $z = 0$ and $\sigma_a = \gamma H$ at $z = H$. The pressure distribution shows that at z=0 the active pressure equals $-2c\sqrt{K_a}$ indicating tensile stress. This tensile stress decreases with depth and becomes zero at a depth $z = z_c$, or

$$\gamma z_c K_a - 2c\sqrt{K_a} = 0$$

And

$$z_c = \frac{2_c}{\gamma\sqrt{K_a}}$$

The depth z_c is usually referred to as the depth of tensile crack, because the tensile stress in the soil will eventually cause a crack along the soil-wall interface. Thus the total Rankine active force per unit length of the wall before the tensile crack occurs is,

$$P_a = \int_0^H \sigma_a dz = \int_0^H \gamma z K_a dz - \int_0^H 2c\sqrt{K_a} dz$$

$$\frac{1}{2}\gamma H^2 K_a - 2cH\sqrt{K_a}$$

Table: Variation of Rankine K_a

Soil friction angle, ϕ (deg)	$Ka = \tan^2(45 - \phi/2)$
20	0.490
21	0.472
22	0.455
23	0.438
24	0.422
25	0.406
26	0.395
27	0.376
28	0.361
29	0.347
30	0.333

31	0.320
32	0.307
33	0.295
34	0.283
35	0.271
36	0.260
37	0.249
38	0.238
39	0.228
40	0.217
41	0.208
42	0.198
43	0.189
44	0.180
45	0.172

After the occurrence of the tensile crack, the force on the wall will be caused only by the pressure distribution between depths $z=z_c$ and $z=H$, as shown by the hatched area in figure above. It may be expressed as,

$$P_a = \frac{1}{2}(H - z_c)\left(\gamma H K_a - 2c\sqrt{K_a}\right)$$

Or

$$P_a = \frac{1}{2}\left(H - \frac{2_c}{\gamma\sqrt{K_a}}\right)\left(\gamma H K_a - 2c\sqrt{K_a}\right)$$

For calculation purposes in some retaining wall design problems, a cohesive soil backfill is replaced by an assumed granular soil with a triangular Rankine active pressure diagram with $\sigma_a = 0$ at $z = 0$ and $\sigma_a = \sigma_v K_a - 2c\sqrt{K_a}$ at $z = H$. In such a case, the assumed active force per unit length of the wall is.

$$P_a = \frac{1}{2}\left(\gamma H K_a - 2c\sqrt{K_a}\right) = \frac{1}{2}\gamma H^2 K_a - cH\sqrt{K_a}$$

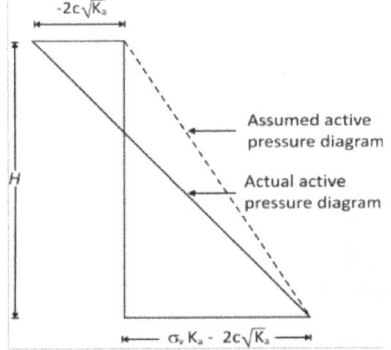

Figure: Assumed active pressure diagram for clay backfill behind a retaining wall

However, the active earth pressure condition will be reached only if the wall is allowed to "yield" sufficiently. The amount of outward displacement of the wall necessary is about 0.001H to 0.004H for granular soil backfills and about 0.01H to 0.04H for cohesive soil backfills.

A 6-m-high retaining wall is to support a soil with unit weight $\gamma = 17.4\,KN/m^3$, soil friction angle $\phi = 26°$ and cohesion c= 14.36 kN/m2. Determine the Rankine active force per unit length of the wall both before and after the tensile crack occurs, and determine the line of action of the resultant in both cases.

For $\phi = 26°$

$$K_a = \tan^2\left(45 - \frac{\phi}{2}\right) = \tan^2(43 - 13) = 0.39$$

$$\sqrt{K_a} = 0.625$$

$$\sigma_a = \gamma H K_a - 2c\sqrt{K_a}$$

Refer to figure(c) above,

At

$$z = 0, \sigma_a = -2c\sqrt{K_a} = -2(14.36)(0.625) = -17.95KN/m^2$$

At

$$z = 6\,m, \sigma_a = (17.4)(6)(0.39) - 2(14.36)(0.625) = 40.72 - 17.95 = 22.77\,KN/m^2$$

Rankine Active Earth Pressure for Inclined Backfill

If the backfill of a frictionless retaining wall is a granular soil $(c = o)$ and rises at an angle a with respect to the horizontal, the active earth pressure coefficient, K_a, may be expressed in the form.

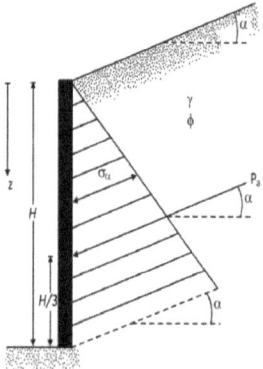

Figure: Notations for active pressure-equations

$$k_a = \cos\alpha\frac{\cos\alpha - \sqrt{\cos^2\alpha - \cos^2\phi}}{\cos\alpha + \sqrt{\cos^2\alpha - \cos^2\phi}}$$

Where

ϕ =angle of friction of soil

At any depth, z, the Rankine active pressure may be expressed as,

$$\sigma_a = \gamma z K_a$$

Also, the total force per unit length of the wall is,

$$P_a = \frac{1}{2}\gamma H^2 K_a$$

Note that, in this case, the direction of the resultant force, P_a , is inclined at an angle α with the horizontal and intersects the wall at a distance of $H/3$ from the base of the wall. Table below presents the values of K_a (active earth pressure) for various values of α and ϕ .

The preceding analysis can be extended for an inclined backfill with a $c-\phi$ soil. As in equation, for this case,

$$\sigma_a = \gamma z K_a = \gamma z K'_a \cos\alpha$$

Table: Active Earth Pressure Coefficient, K_a

	$\phi\,(\text{deg}) \rightarrow$						
$\downarrow\alpha\,(\text{deg})$	28	30	32	34	36	38	40
0	0.361	0.333	0.307	0.283	0.260	0.238	0.217
5	0.366	0.337	0.311	0.286	0.262	0.240	0.219
10	0.380	0.350	0.321	0.294	0.270	0.246	0.225
15	0.409	0.373	0.341	0.311	0.283	0.258	0.235
20	0.461	0.414	0.374	0.338	0.306	0.277	0.250
25	0.573	0.494	0.434	0.385	0.343	0.307	0.275

Table: Values of K_a

$\phi\,(\text{deg})$	$\alpha\,(\text{deg})$	$\dfrac{c}{\gamma z}$			
		0.025	-0.05	0.1	0.5
15	0	0.550	0.512	0.435	-0.179
	5	0.566	0.525	0.445	-0.184
	10	0.621	0.571	0.477	-0.186
	15	0.776	0.683	0.546	-0.196
20	0	0.455	0.420	0.350	-0.210
	5	0.465	0.429	0.357	-0.212
	10	0.497	0.456	0.377	-0.218
	15	0.567	0.514	0.417	-0.229
25	0	0.374	0.342	0.278	-0.231

	5	0.381	0.348	0.283	-0.233
	10	0.402	0.366	0.296	-0.239
	15	0.443	0.401	0.321	-0.250
30	0	0.305	0.276	0.218	-0.244
	5	0.309	0.280	0.221	-0.246
	10	0.323	0.292	0.230	-0.252
	15	0.350	0.315	0.246	-0.263

Where,

$$K'_a =$$

$$\frac{1}{\cos^2 \phi} \left\{ \frac{2\cos^2 \alpha + 2\left(\dfrac{c}{yz}\right)\cos\phi\sin\phi}{-\sqrt{[4\cos^2 \alpha\left(\cos^2 \alpha - \cos^2 \phi\right) + 4\left(\dfrac{c}{yz}\right)^2 \cos^2 \phi + \left(\dfrac{c}{yz}\right)\cos^2 \alpha + 8\left(\dfrac{c}{yz}\right)\cos^2 \alpha \sin\phi\cos\phi]}} \right\} - 1$$

Some values of K'_a are given in table above. For a problem of this type, the depth of tensile crack, z_c, is given as,

$$z_c = \frac{2c}{\gamma}\sqrt{\frac{1+\sin\phi}{1-\sin\phi}}$$

Given: $H = 7.5m$, $\gamma = 18\,KN/m^3$, $\phi = 20°$, $c = 13.5\,KN/m^2$, and $\alpha = 10°$. Calculate the Rankine active force, P_a, per unit length of the wall and the location of the resultant after the occurrence of the tensile crack.

Solution

From equation above

$$z_c = \frac{2c}{\gamma}\sqrt{\frac{1+\sin\phi}{1-\sin\phi}} = \frac{(2)(13.5)}{18}\sqrt{\frac{1+\sin 20}{1-\sin 20}} = 2.14\,m$$

At $z = 7.5\,m$

$$\frac{c}{yz} = \frac{13.5}{(18)(7.5)} = 0.1$$

From table above, for $20°, c/yz = 0.1$ and $\alpha = 10°$, the value of K'_a is

$$0.377, \text{ so at } z = 7.5\,m\,\sigma_a = \gamma z K'_a \cos\alpha = (18)(7.5)(0.377)(\cos 10) = 50.1\,KN/m^2 .$$

After the occurrence of the tensile crack, the pressure distribution on the wall will be as shown in figure below, so

$$P_a = \left(\frac{1}{2}\right)(50.1)(7.5 - 2.14) = 134.3KN/m$$

$$\bar{z} = \frac{7.5 - 2.14}{3} = 1.79\,m$$

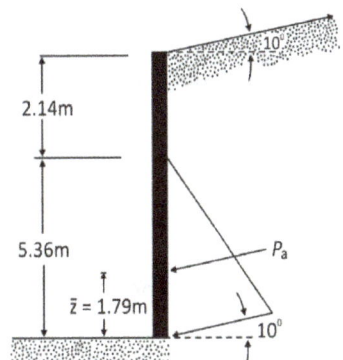

Terzaghi's Effective Stress Principle

The total stress increases when additional vertical load is first applied. Instantaneously, the pore water pressure increases by exactly the same amount. Subsequently there will be flow from regions of higher excess pore pressure to regions of lower excess pore pressure causing dissipation. The effective stress will change and the soil will consolidate with time. This is shown schematically.

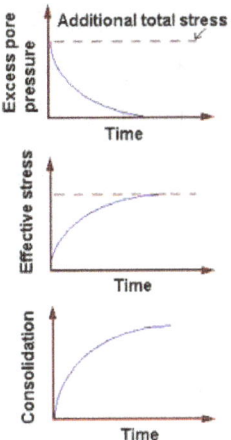

On the assumption that the excess pore water drains only along vertical lines, an analytical procedure can be developed for computing the rate of consolidation.

Consider a saturated soil element of sides dx, dy and dz.

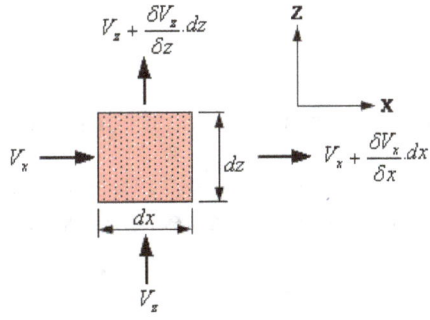

The initial volume of soil element = dx.dy.dz

If n is the porosity, the volume of water in the element = n.dx.dy.dz

The continuity equation for one-dimensional flow in the vertical direction is,

$$\frac{\delta Vz}{\delta z}dx.dy.dz = -\frac{\delta}{\delta t}(n.dx.dy.dz)$$

Only the excess head (h) causes consolidation, and it is related to the excess pore water pressure (u) by $h = u/g_w$. The Darcy equation can be written as,

$$V_z = -k_z\frac{\delta h}{\delta z} = -\frac{k_z}{\gamma_w}\frac{\delta h}{\delta z}$$

The Darcy eqn. can be substituted in the continuity eqn., and the porosity n can be expressed in terms of void ratio e, to obtain the flow equation as,

$$\frac{k_z}{\gamma_w}\frac{\delta^2 u}{\delta z^2}dx.dy.dz = \frac{\delta}{\delta t}\left(\frac{e}{1+e}dx.dy.dz\right)$$

The soil element can be represented schematically as,

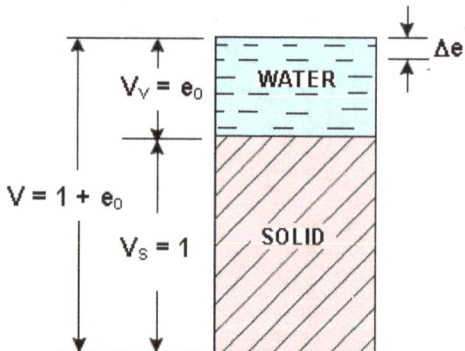

If e_o is the initial void ratio of the consolidating layer, the initial volume of solids in the element is (dx dy dz) / (1 + e_o), which remains constant. The change in water volume can be represented by small changes De in the current void ratio e.

The flow eqn. can then be written as,

$$\frac{k_z}{\gamma_w}\frac{\delta^2 u}{\delta z^2}dx.dy.dz = \frac{dx.dy.dz}{1+e_0}.\frac{\delta e}{\delta t}$$

or

$$\frac{k_z}{\gamma_w}\frac{\delta^2 u}{\delta z^2} = \frac{1}{1+e_0}\frac{\delta t}{\delta t}$$

This is the hydrodynamic equation of one-dimensional consolidation.

If a_v = coefficient of compressibility, the change in void ratio can be expressed $\Delta e = a_v.(-\Delta\sigma') = a_v.(\Delta u)$ since any increase in effective stress equals the decrease in excess pore water pressure. Thus,

$$\frac{\delta e}{\delta t} = a_v\frac{\delta u}{\delta t}$$

The flow eqn. can then be expressed as,

$$\frac{k_z}{\gamma_w}\frac{\delta^2 u}{\delta z^2}=\frac{a_v}{1+e_0}\cdot\frac{\delta u}{\delta t}$$

or

$$\frac{k_z}{a_v}\cdot\frac{(1+e_0)}{\gamma_w}\cdot\frac{\delta^2 u}{\delta z^2}=\frac{\delta u}{\delta t}$$

By introducing a parameter called the coefficient of consolidation, $c_v=\dfrac{k_z(1+e_0)}{a_v\cdot\gamma_w}=\dfrac{k_z}{m_v\cdot\gamma_w}$ the flow eqn. then becomes,

$$c_v=\frac{\delta^2 u}{\delta z^2}\frac{\delta u}{\delta t}$$

This is Terzaghi's one-dimensional consolidation equation. A solution of this for a set of boundary conditions will describe how the excess pore water pressure u dissipates with time t and location z. When all the u has dissipated completely throughout the depth of the compressible soil layer, consolidation is complete and the transient flow situation ceases to exist.

Solution of Terzaghi's Theory

During the consolidation process, the following are assumed to be constant:

1. The total additional stress on the compressible soil layer is assumed to remain constant.

2. The coefficient of volume compressibility (mV) of the soil is assumed to be constant.

3. The coefficient of permeability (k) for vertical flow is assumed to be constant.

There are three variables in the consolidation equation:

1. the depth of the soil element in the layer (z)

2. the excess pore water pressure (u)

3. the time elapsed since application of the loading (t)

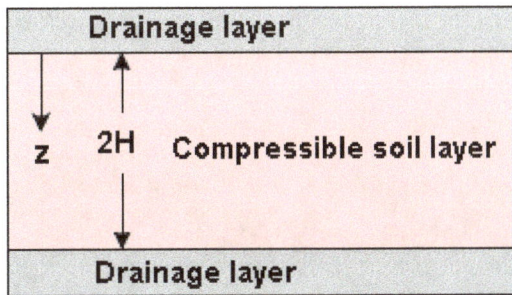

To take care of these three variables, three non-dimensional parameters are provided:

1. Drainage path ratio, $Z=\dfrac{z}{H}$, where H = drainage path which is the longest path taken by the pore water to reach a permeable sub-surface layer above or below.

2. Consolidation ratio at depth z, U_z, which is the ratio of dissipated pore pressure to the initial excess pore pressure. This represents the stage of consolidation at a certain location in the compressible layer.

3. Time factor, $T = \dfrac{c_v.t}{H^2}$

The graphical solution of Terzaghi's one-dimensional consolidation equation using the non-dimensional parameters is shown.

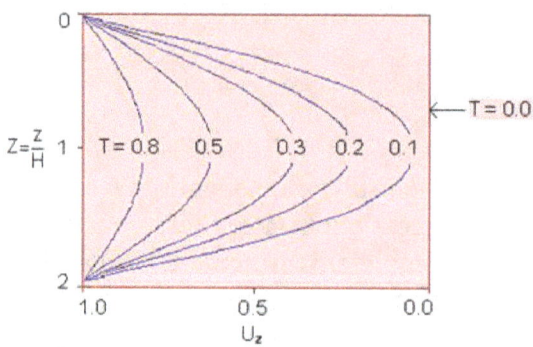

The figure is symmetrical about the horizontal line at $Z = \dfrac{z}{H} = 1$

For double drainage conditions, pore water above this location flows upwards whereas water below this location flows downwards. Thus, the horizontal line at $Z = 1$ is equivalent to an imperious boundary. For single drainage conditions, only either the top half or bottom half of the figure is to be used, and the drainage path is equal to the thickness of the compressible layer.

The above graphical solution shows how consolidation proceeds with time at different locations for a particular set of boundary conditions, but it does not describe how much consolidation occurs as a whole in the entire compressible layer.

The variation of total consolidation with time is most conveniently plotted in the form of the average degree of consolidation (U) for the entire stratum versus dimensionless time T, and this is illustrated below.

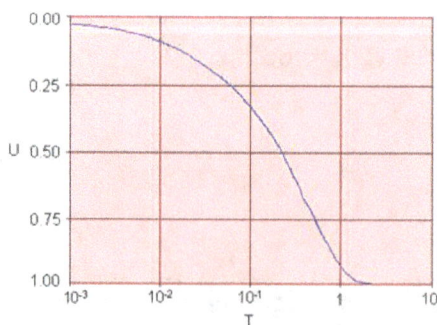

There are useful approximations relating the degree of consolidation and the time factor, viz:

For $U \le 0.60$, $T = (\pi/4).U^2$

For $U > 0.60$, $T = 1.781 - 0.933 \log_{10}(100 - U\%)$

Consolidation Settlement and Time

To estimate the amount of consolidation which would occur and the time it would take to occur, it is necessary to know:

1. The boundary and drainage conditions;

2. The loading conditions;

3. The relevant parameters of the soil, including initial void ratio, coefficient of compressibility, coefficient of volume compressibility, compression index, and coefficient of consolidation. They are obtained from consolidation tests on representative undisturbed samples of the compressible soil stratum.

ΔD = Change in thickness

D = Thickness of compressible soil layer

Comparing the compressible soil layer with a soil element of this layer,

$$\frac{Change\,in\,thickness}{Total\,thickness} = \frac{Change\,in\,volume}{Total\,volume}$$

$$\frac{\Delta D}{D} = \frac{\Delta e}{1+e_o}$$

$$\therefore \Delta D = \frac{\Delta e}{1+e_o} D$$

Δe can be expressed in terms of a_v or C_c.

$$\Delta e = a_v \Delta \sigma' \quad \text{Or} \quad \Delta e = C_c \log(\sigma'_0 + \Delta\sigma'/\sigma'_0)$$

Or

$$\Delta D = \frac{a_v . \Delta\sigma}{1+e_o} . D = m_v . \Delta\sigma . D$$

Or,

$$\Delta D = \frac{C_c \log\left(\dfrac{\sigma_o + \Delta\sigma}{\sigma_o}\right) . D}{1+e_o}$$

Split Spoon Sampler

When taking samples, one of the more popular options in sampling equipment is the split spoon. Split spoon sampling is useful in sampling for lithological descriptions, geotechnical analysis that does not need undisturbed sampling, and for chemical analysis.

The split spoon sampler is a tube split into two equal halves lengthwise. The two halves are locked together during the sampling activities and released to retrieve the samples. At bottom end of the sampler sits a driving shoe. This is what cuts into the soil and provides the sample that goes up into the tube. At the other end of the tube is a coupling that allows it to connect to the drilling rod. Once a sample is taken, the operator removes the ends from the tube. This allows the tube to "split" open. Representative samples can be taken and put into containers for shipping to a lab for analysis, observed and field tested in the field, or discarded in a proper waste container if not warranted for the investigation.

SPT test is performed to determine some properties of soils, especially in cohesion less soils, for which undisturbed samples cannot be simply obtained. The SPT utilizes split spoon sampler. It is a tube of dimensions equal to 51 mm outer diameter and 35 mm inner diameter. It is 457-610 mm in long that is split down the middle longitudinally.

Sampler will be connected to bottommost part in the drilling rod. The drilling rod is pushed into soil by using drop hammer. The hammer with 140 N weight falling from the height of 762 mm is employed to drive the sampler to the depth of 457 mm into soil.

The SPT setup is as shown in the below figure.

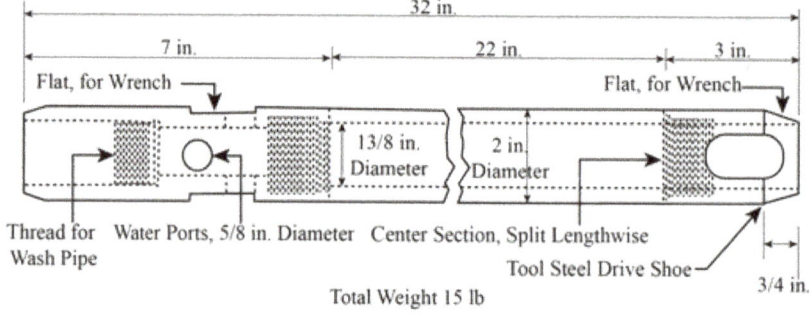

Split-spoon sampler for the sandard penetration test.

The blows used to drive each of three 152 mm increments are recorded individually. The value of total blows used for the last 305 mm is known as Standard Penetration Resistance value. Thus, the N-value denotes the blows required per mm depth. When the blow counts are completely recorded, the sampler can be taken out. Then, it is completely opened to get a disturbed sample for consequent testing and examination.

Overburden pressure influences the SPT results at the place, where the blow counts were made. Numerous methods are used to correct the N-values based on the over burden pressure influence. Two methods are presented as below.

One method utilizes the following equation to evaluate CN, a correction factor N' for the N-value at field.

$$C_N = 0.77 \log_{10} \frac{1915}{p_o}$$

Here, the value of effective overburden pressure at the elevation of the SPT is termed as po.

Uses of Split Spoon Sampling

One of the unique advantages of this kind of sampling equipment is that it can be used on most drilling rigs. The drilling rigs will drill down to a certain level and, at that point, the drilling team will install the split spoon and insert it down to the bottom of the hole. However if you are using hollow stem auger drilling methods you can simultaneously drive a split spoon while drilling. Using a drilling hammer, the split spoon gets driven down into the soil, and then pulled out. This kind of sampling works best in soft to compacted soils as long as there is not a large amount of rocks or cobbles.

Core Catcher and its use

Some soils are silty or sandy and will not stay in the split spoon as it is drawn up. To prevent this from happening, a core catcher can be used. The catcher allows soil to enter the sampler, but prevents it from exiting. When the tube is withdrawn from the sampling hole, the soil will stay in place until the catcher is removed or the tube is split.

Use a Sleeve in a Split Spoon Sampler

When samples are being taken for chemical analysis, it is important to encase them in an inert material. To do this, a driller will insert stainless, carbon, brass, or acetate sleeves into the split spoon. Once the sample is taken, one of two things can happen. The entire sleeve gets capped and it goes to the lab for testing. Or the contents of the sleeve get emptied into an inert container for shipping to the lab.

Standard Penetration Test

The standard penetration test is an in-situ test that is coming under the category of penetrometer tests. The standard penetration tests are carried out in borehole. The test will measure the resistance of the soil strata to the penetration undergone. A penetration emphirical correlation is derived between the soil properties and the penetration resistance.

The test is extremely useful for determining the relative density and the angle of shearing resistance of cohesion less soils. It can also be used to determine the unconfined compressive strength of cohesive soils.

Tools for Standard Penetration Test

The requirements to conduct SPT are:

1. Standard Split Spoon Sampler.

2. Drop Hammer weighing 63.5kg.

3. Guiding rod.

4. Drilling Rig.

5. Driving head (anvil).

Procedure for Standard Penetration Test

The test is conducted in a bore hole by means of a standard split spoon sampler. Once the drilling is done to the desired depth, the drilling tool is removed and the sampler is placed inside the bore hole.

By means of a drop hammer of 63.5 kg mass falling through a height of 750 mm at the rate of 30 blows per minute, the sampler is driven into the soil.

The number of blows of hammer required to drive a depth of 150 mm is counted. Further it is driven by 150 mm and the blows are counted.

Similarly, the sampler is once again further driven by 150mm and the number of blows recorded. The number of blows recorded for the first 150mm not taken into consideration. The number of blows recorded for last two 150 mm intervals are added to give the standard penetration number (N). In other words.

N = No: of Blows Required for 150 mm Penetration Beyond Seating Drive of 150 mm.

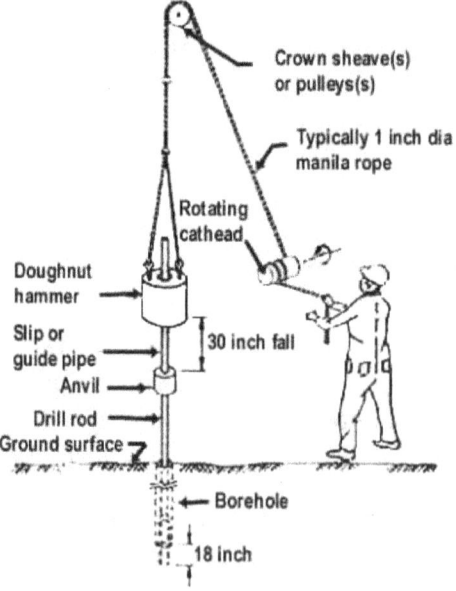

Figure: Standard penetration Test

If the number of blows for 150mm drive exceeds 50, it is taken as refusal and the test is discontinued. The standard penetration number is corrected for dilatancy correction and overburden correction.

Corrections in Standard Penetration Test

Before the SPT values are used in empirical correlations and in design charts, the field 'N' value have to be corrected as per IS 2131 – 1981. The corrections are:

1. Dilatancy Correction

2. Overburden Pressure Correction

Dilatancy Correction

Silty fine sands and fine sands below the water table develop pore water pressure which is not easily dissipated. The pore pressure increases the resistance of the soil and hence the penetration number (N).

Terzaghi and Peck recommend the following correction in the case of silty fine sands when the observed value is N exceeds 15.

The corrected penetration number,

$$N_C = 15 + 0.5 (N_R - 15)$$

Where N_R is the recorded value and N_C is the corrected value.

If N_R less than or equal to 15, then $N_c = N_R$

Overburden Pressure Correction

From several investigations, it is proven that the penetration resistance or the value of N is dependent on the overburden pressure. If there are two granular soils with relative density same, higher 'N' value will be shown by the soil with higher confining pressure.

With the increase in the depth of the soil, the confining pressure also increases. So the value of 'N' at shallow depth and larger depths are underestimated and overestimated respectively.

Hence, to account this the value of 'N' obtained from the test are corrected to a standard effective overburden pressure.

The corrected value of 'N' is

$$Nc = C_N N$$

Here CN is the correction factor for the overburden pressure.

Precautions taken for Standard Penetration Test

1. Split spoon sampler must be in good condition.

2. The cutting shoe must be free from wear and tear.

3. The height of fall must be 750 mm. Any change from this will affect the 'N' value.

4. The drill rods used must be in standard condition. Bent drill rods are not used.

5. Before conducting the test, the bottom of the borehole must be cleaned.

Advantages of Standard Penetration Test

The advantages of standard penetration test are:

1. The test is simple and economical.

2. The test provides representative samples for visual inspection, classification tests and for moisture content.

3. Actual soil behavior is obtained through SPT values.

4. The method helps to penetrate dense layers and fills.

5. Test can be applied for variety of soil conditions.

Disadvantages of Standard Penetration Test

The limitations of standard penetration tests are:

1. The results will vary due to any mechanical or operator variability or drilling disturbances.

2. Test is costly and time consuming.

3. The samples retrieved for testing is disturbed.

4. The test results from SPT cannot be reproduced.

5. The application of SPT in gravels, cobbles and cohesive soils are limited.

Cone Penetration Test

A cone penetration test is used to determine geotechnical properties of soils. The cone penetration test has become internationally one of the most widely used and accepted test methods for determining geotechnical soil properties. the data gained from a cone penetration test can be used to assess whether soil layers are likely to liquefy under different levels of earthquake shaking.

Cone Penetration Test Undertaken

The cone penetration test can be completed from the ground surface. Cone penetration test rigs vary in size – from small portable rigs to large truck-mounted rigs. Each rig has benefits and limitations but they all conduct the same test. A cone penetration test rig pushes a steel cone (about 32 mm wide) into the ground, generally up to 20 m below the surface or until the cone reaches a hard layer. The steel cone contains an electronic measuring system that records tip resistance and sleeve friction.

As the cone is pushed into the ground, the soil responds with differing degrees of resistance. This resistance is recorded using force sensors in the tip.

At the same time as the sensors are recording resistance at the cone tip, sensors in the friction sleeve are recording sleeve friction along a 100mm length. Some cones also have a pore water transducer, which records water pressure in the soil. These readings can be used to determine ground water responses as the cone is pushed through the soils.

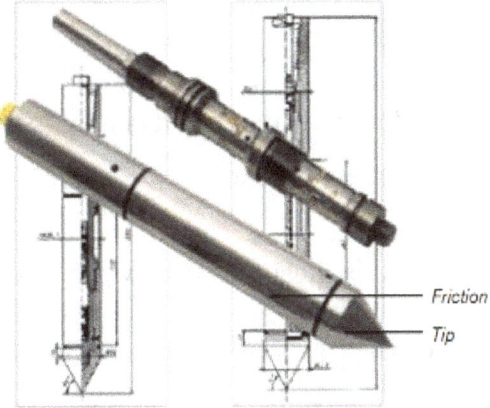
Friction
Tip

A cone penetration test typically takes between 30 minutes and three hours. As the cone goes into the ground, measurements are constantly sent back to the rig and recorded on computer.

This data gives a profile of the subsoil layers, often called a 'trace'. Examples are shown to the right.

Test Results

Cone penetration test results are used by geotechnical engineering specialists to understand the soil properties (the relative density of the soil and the soil behavior type, both of which are calculated from the cone penetration test cone tip resistance and sleeve friction) and how the ground is likely to behave under different levels of earthquake shaking. This information can help in the design of foundations and ground improvements works.

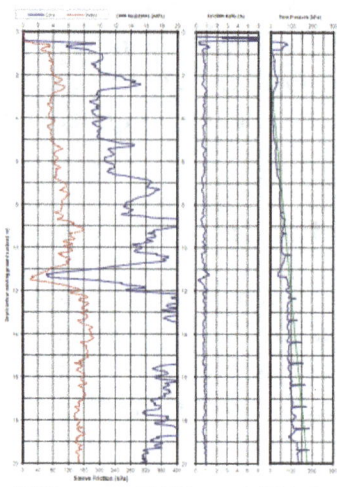

Cone penetration test results are commonly used to determine the liquefaction-triggering resistance of each soil layer. These assessments commonly use computer software to determine if soil layers are predicted to liquefy for different levels of earthquake shaking.

By doing a test before and after ground improvement works, cone penetration test results can also be used to determine how much strength a soil has gained following ground improvement works.

Benefits of Cone Penetration Test

Cone penetration testing can provide a wide range of information in suitable soils, delivering large volumes of data to internationally recognized formats at lower cost and with less impact than alternative methods. CPTs have the following features and benefits:

- Low impact: CPT provides a continuous subsurface picture with less ground disturbance than drilling.

- Highly mobile: CPT can be used in most situations where ground investigation is required. On land, it can be deployed from trucks, tracked vehicles and bespoke vehicles, such as rail wagons. Over water, CPTs can be conducted from vessels, barges and jack-up platforms.

- In situ measurement: therefore less affected by sample disturbance or stress relief than laboratory-based methods.

- Versatility: we offer modular systems with many types of sensor available, to yield a range of information from a single mobilization.

- Big data, delivered quickly: with rapid acquisition (>100 m per day), CPT yields thousands of data points to enable reliable site characterization, and provides digital output that can be transferred in real time.

- Widely recognized: CPT outputs are internationally recognized, meeting most commonly used design codes and quality standards.

- Always evolving: over the course of our 50 years' CPT development, we have continually worked to improve CPT safety, resilience and technical performance.

Vane Shear Test

Vane shear test is used to determine the undrained shear strength of soils especially soft clays. This test can be done in laboratory or in the field directly on the ground. Vane shear test gives accurate results for soils of low shear strength (less than 0.3 kg/cm²).

Shear Strength of Soil by Vane Shear Test

Apparatus

Apparatus required for vane shear test are:

1. Vane shear apparatus

2. Soil specimen container

3. Vernier calipers

Vane shear apparatus consists high tensile steel rod to which four steel blades (vanes) are fixed at right angles to each other at the bottom of rod.

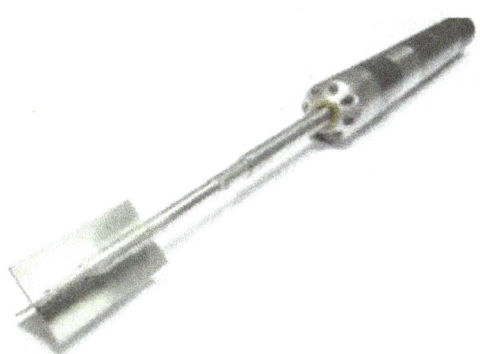

Procedure of Vane Shear Test

Test procedure of vane shear test contains following steps:

1. Clean the vane shear apparatus and apply grease to the lead screw for better movement of handles.

2. Take the soil specimen in container which is generally 75 mm in height and 37.5 mm in diameter.

3. Level the soil surface on the top and mount the container on the base of vane shear test apparatus using screws provided.

4. Lower the vane gradually into the soil specimen until the top of vane is at a depth of 10 to 20 mm below the top of soil specimen.

5. Note down the reading of pointer on circular graduated scale which is initial reading.

6. Rotate the vane inside the soil specimen using torque applying handle at a rate of 0.10 per second.

7. When the specimen fails, the strain indicator pointer will move backwards on the circular graduated scale and at this point stop the test and note down the final reading of pointer.

8. The difference between Initial and final readings is nothing but the angle of torque.

9. Repeat the procedure on two more soil specimens and calculate the average shear strength value.

10. Measure the diameter and height of vane using Vernier callipers.

11. Sensitivity of given soil sample is determined by repeating the above test procedure on re-molded soil which is nothing but soil obtained after rapid stirring of vane in the above test.

Sensitivity of soil = undisturbed shear strength/ remolded shear strength.

Advantages of Vane Shear Test

Advantages of vane shear test are as follows:

*   Vane shear test is easy and quick.

- This test can be performed either in laboratory or in the field directly on the ground.

- In-situ vane shear test ideal for the determination of undrained shear strength of non-fissured, fully saturated clay.

- Shear strength of soft clays at greater depths can also be found by vane shear test.

- Sensitivity of soil can also be determined using vane shear test results of undisturbed and remolded soil samples.

Drawbacks of Vane Shear Test

Drawbacks of vane shear test are as follows:

- Vane shear test is not suitable for clays which contain sand or silt laminations in it.

- It cannot be conducted on the fissured clay.

- If the failure envelope is not horizontal, vane shear test does not give accurate results.

References

- Soil, bearing-capacity-of-soil-with-diagram-45824: yourarticlelibrary.com, Retrieved 28 May 2018

- Soil-classification-and-identification-with-diagram-45407: yourarticlelibrary.com, Retrieved 12 June 2018

- The-proper-use-of-a-split-spoon-sampling-device: talonlpe.com, Retrieved 27 April 2018

- Split-spoon-sampler-8: chegg.com, Retrieved 13 July 2018

- Standard-penetration-test-procedure-precautions-advantages-4657: theconstructor.org, Retrieved 06 July 2018

- Geotechnical, vane-shear-test-on-soil-3435: theconstructor.org, Retrieved 13 June 2018

Rock Mechanics

Rock mechanics is an applied and theoretical science concerned with the study of the mechanical behavior of rocks, such as the influence of force fields on rock masses. An important technique for rock mass classification and rock slope engineering is the Q-slope method. All the diverse principles of rock mechanics and Q-slope have been carefully examined in this chapter.

Rock mechanics is the theoretical and applied science of the mechanical behavior of rock. It is that branch of mechanics concerned with the response of rock to the force fields of its physical environment."

It is field of study that overlaps into other sciences such as geology, soil mechanics, and geophysics. The fundamentals of rock mechanics include rock behavior, rock mass response, flow, chemistry, and coupled behavior. Some of these, such as rock mass response, can be simulated in a laboratory or studied through theoretical modeling of field-based data.

Deformation of Rocks

When rocks are under stress, they can become deformed by fracturing. There are three types of stress that can cause deformation. These include stretching (tensional stress), compression (compressional stress), and slippage (shear stress). Different types of rocks respond to stressors in different ways.

Clay-based formations, for example, have a high degree of elasticity and can be reshaped under pressure without breaking apart. On the other hand, rocks such as quartz and feldspar are extremely brittle and can easily fracture when stressed.

Brittle fractures are explained through The Griffith Theory, which suggests that inherent cracks in the rock formation, when put under pressure, can result in a phenomenon called "rockburst." This is an occurrence characteristic of excavation sites and in instances of severe compressional stress.

Skill Sets that are Applied to the Study of Rock Mechanics

There are several competencies that come into play in rock mechanics. Some of these include lab testing, field testing, risk analysis, site management, equipment development, numerical modeling, and information technology. Current research into rock mechanics utilizes a combination of skill sets to advance the study of:

- Effect of time and fluids on rock formations.

- Static and dynamic behavior of rocks at the crystal level.

- Long-term behavior of tunnels and other underground structures.

- Advanced simulation methodologies using parallel computing.

- Alternative measurement strategies for acoustic emissions, magnetic fields, and more.

- Improved digital image techniques for better pattern recognition.

- Deeper exploration into fault zones, ocean foundations, and outer space.

- Predictability of earthquake triggers and similar geological phenomenon.

Industries that Utilize Rock Mechanics

There are a number of industries that employ engineers with knowledge of rock mechanics. The mining, petroleum and construction industries are at the top of the list. Construction has multiple applications in infrastructure, tunneling, irrigation, dam building, and more. It is expected that demand for rock mechanics specialists will increase due to global trends such as:

- Demand for materials found in hard-to-mine locations.

- Growth in infrastructure and hydropower construction in developing countries.

- The need to secure unsafe rock formations in populated areas.

- Investment in geothermal energy and hydrocarbons storage.

It's not hard to understand the depth of importance of rock mechanics as an essential part of achieving success in a variety of civil engineering career areas including transportation, wastewater, underground construction and more.

Problems that Arise in Rock Mechanics

Civil engineers have to routinely deal with geotechnical matters where natural conditions remain unknown and inferences have to be made based on observations and experience, with some assistance from laboratory testing. By contrast, the applied science of mechanics and structural engineering is based on deduction that gives definite results. These two aspects have to be considered when you try to understand what rock mechanics is and where an engineer has to assess the properties and strengths of the rock that he can use for foundations for structures.

Rock mechanics determines how a particular rock reacts when it is put to the use required by mankind for buildings, roads, bridges, other civil engineering uses. It will assess the bearing capacity of the rock on the surface and how the force applied on the rock by the structures being built on it will affect the rock at various depths. Rock mechanics will determine the shear strength of the rock, which in turn will allow the rock to resist the forces applied to it. Rock mechanics can also determine the response of rock when it is subjected to dynamic loading that may be a result of manmade applications or natural occurrences like earthquakes. The failure mechanism of rocks will allow engineers to counteract these so that the structures built on the rock are safe. Rock mechanics will also study the effect that defects in the rock from cavities, fissures, joints and bedding planes can have on structures founded on them.

Rock mechanics will also allow engineers to decide how to protect slopes, the proper technique to be used for tunneling, the strengths that can be expected from rock that functions as ballast for railway tracks or as base for roads. The strength of rock also plays a large part in aggregate used for concrete that makes up most of the buildings being built nowadays.

Q-slope

In both civil engineering and mining projects, it is practi- cally impossible to assess the stability of rock slope cuttings and benches in real time, using analytical approaches such as kinematics, limit equilibrium, or FEM/DEM modeling. Excavation is usually too fast for this. The same limitation usually applies to tunneling, despite numerical modeler's wishes to the contrary. However, rock caverns of larger span are sufficiently 'stationary' for thorough and more necessary analysis, and the same applies to higher rock slopes. The purpose of Q-slope is to allow engineering geologists and rock engineers to assess the stability of excavated rock slopes in the field and make potential adjustments to slope angles as rock mass conditions become visible during construction. Prime areas of application are 'from surface and down wards' bench angle decisions in open pits, and for the numerous slope cuttings needed to reach remote hydro- power projects, tunnel and dam sites, often through strongly varying structural geologies. Tens, if not hundreds of thousands of kilometers of such rock cuttings, without support, are made each year. A significant proportion may fail during construction, and shallower slope angles are then chosen. Application of Q-slope can help to reduce maintenance (and bench-width needs) due to all the potential failures. Such are frequently seen when initially constant slope angles are cut through varied structural domains. A series of 'interesting' but troublesome local failures is often the result. Quite often these have been the result of adverse plane failures, wedge failures, or more rarely local toppling. Figure illustrates the very basic classes of potential behavior. These may be accentuated when one of the illustrated joint sets is dominant. Anisotropic behavior is typical, as emphasized by Barton and Quadros. Several combinations of slope angles and jointing in figure are expected to be stable (i.e., a2, b3, c3, which are [45°; and d1, d2, d3, d4, which are equal to 45° in the idealized examples illustrated).

Joint or discontinuity shear strength obviously has a significant role in rock slope stability. On a more sophisticated level than in figure, it may often be a coupled problem where shear-deformation can improve drainage temporarily due to dilation (figure), until fractures are clogged with run-off fines in future storms. Therefore, the environment in which the slopes are constructed is also critical. The sheared and dilated tension- fracture replicas on the right are from Barton, and the three potential joint behavior modes involving mobilized shear strength, dilation, and increase of permeability are based on coupled M–H (mechanical–hydraulic) Barton–Bandis joint modeling, as described in Barton et al.

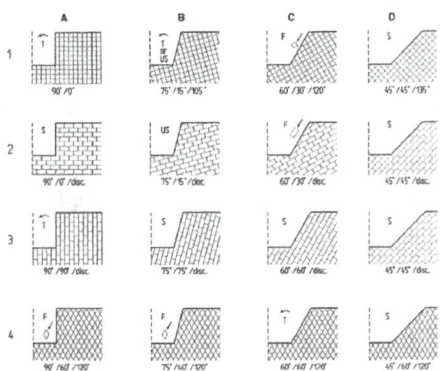

Common sense solutions to dominant joint orientation problems
(*S* stable, *F* sliding failure, *T* toppling, *US* unstable)

For Q-system users, the formula for estimating Q- slope is mostly familiar:

The Q-Slope Method

The Q-slope method requires the assignment of ratings for rock quality designation (RQD), joint set number (J_n), joint roughness number (J_r), and joint alteration number (J_a), which remain unchanged from the Q-system. For Q-system users, the formula for estimating Q- slope is mostly familiar:

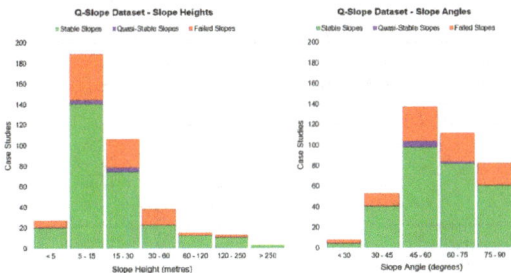

Q-slope dataset (case studies)—slope heights (*left*) and slope angles (*right*)

$$Qslope = \frac{RQD}{J_n} \times \left(\frac{J_r}{J_a}\right)_0 \times \frac{J_{wice}}{SRF_{slope}}$$

As with the Q-system, the rock mass quality in Q-slope can be considered a function of three parameters, which are crude measures of:

1. Block size: (RQD/J_n).

2. Shear strength: least favorable (J_r/J_a) or average shear strength in the case of wedges (J_r/J_a)1 × (J_r/J_a)2.

3. External factors and stress: (J_{wice}/SRF_{slope}). Shear resistance, s, is approximated using:

Shear resistance, τ, is approximated using:

$$\tau \approx \sigma\,tan^{-1}\left(\frac{J_r}{J_a}\right)$$

Discontinuity Orientation Factor: O-Factor

The discontinuity orientation factor (O-factor) described in table provides orientation adjustments for discontinu- ities in rock slopes. Figure provides photographic examples.

The Set A orientation-factor is applied to the most unfavorable discontinuity set. If required, the Set B orientation-factor is applied to the secondary discontinuity set (i.e., in case of potentially unstable wedge formations).

Environmental and Geological Conditions: J_{wice}

The environmental and geological condition number, $J_{wice,}$ is more sophisticated than J_w of the original Q-system

Table: Rock quality designation

Rock quality designation descriptionn		RQD(%)[a]
A	Very poor	0–25
B	Poor	25–50
C	Fair	50–75
D	Good	75–90
E	Excellent	90–100

where RQD is reported or measured as B10 (including zero), a nominal value of 10 is used to evaluate Q-slope. RQD intervals of 5, i.e., 100, 95, 90, etc., are sufficiently accurate since slopes are outside and exposed to the elements for a very long time.

Table: Joint set number

Joint set number description		J_n
A	Massive, no or few joints	0.5-1
B	One joint set	2
C	One joint set plus random joints	3
D	Two joint sets	4
E	Two joint sets plus random joints	6
F	Three joint sets	9
G	Three joint sets plus random joints	12
H	Four or more joint sets, random, heavily jointed	15
J	Crushed rock, earthlike	20

Described in table, Jwicehas a new structure forslopes, including tropical rainfall erosion effects and ice-wedging effects. Adjustment factors in case of slope rein-forcement or drainage measures are also included. Competent rocks are generally durable, resistant toerosion and deformation, and not susceptible to slaking. Often these have a relatively high unconfined compressivestrength, say 50 MPa and above. The estimate of Jwiceshould take into consideration the environmental conditions in which the slope is constructed, which will includethe competence or otherwise of the rock, and therefore thelikely long-term stability of possibly adverse structures. The most hostile or dynamic environmental conditionsexperienced by the slope should be adopted. For example,if a slope is constructed in a desert environment that regularly experiences extremely cold winters, freezing, andthawing, it would be appropriate to adopt ice wedging asthe most adverse environmental condition.

Strength Reduction Factor

The strength reduction factor SRFslope is obtained by using the most adverse i.e., maximum of SRFa, SRFb, and SRFc described in the subsequent tables.

Table describes strength reduction factors (SRFa) for the physical condition of the slope surface (now or expected) due to susceptibility to weathering and erosion. Table describes strength reduction factors (SRFb) for adverse stress-strength ranges in the slope. SRFb becomes more critical for weak, low strength materials such as highly weathered and saprolitic rocks, and also becomes more critical with increasing slope height, and therefore, with increasing stress. Maximum principal stress (r1) may be estimated by considering in situ stresses, material density, and slope geometry. However, in the unlikely event that slope-related in situ stress measurements are available, these may be preferred. This might apply in the case of high slopes.

(a) Rock-wall contact, (b) contact after shearing		
A	Discontinuous joints	4
B	Rough or irregular, undulating	3
C	Smooth, undulating	2
D	Slickensided, undulating	1.5
E	Rough or irregular, planar	1.5
F	Smooth, planar	1.0
G	Slickensided, planar	0.5
(c) No rock-wall contact when sheared		
H	Zone containing clay minerals thick enough to prevent rock-wall contact.	1.0
J	Sandy, gravely or crushed zone thick enough to prevent rock-wall contact.	1.0

Descriptions refer to small-scale features and intermediate scale features, in that order Add 1.0 if mean spacing of the relevant joint set is greater than 3 m.

$J_r = 0.5$ can be used for planar, slickensided joints having lineations, provided the lineations are oriented for minimum strength.

J_r and J_a classification are applied to the discontinuity set or sets that are least favorable for stability both from the point of view of orientation and shear resistance s, where s & rn tan-1 (J_r/J_a).

Joint alteration number

Joint alteration number description

		Ja
A	Tightly healed, hard non-softening, impermeable filling, i.e., quartz or epidote	0.75
B	Unaltered joint walls, surface staining only	1.0
C	Slightly altered joint walls. Non-softening mineral coatings, sandy particles, clay-free disintegrated rock, etc.	2.0
D	Silty- or sandy-clay coatings, small clay disintegrated rock, etc. 3.0	
E	Softening or low friction clay mineral coatings, i.e., kaolinite or mica. Also chlorite, talc, gypsum, graphite, etc., and small quantities of swelling clays	4.0
(b)	Rock-wall contact after some shearing (thin clay fillings, probable thickness $\approx 1-5mm$)	
F	Sandy particles, clay-free disintegrated rock, etc.	4.0
G	Strongly over-consolidated non-softening clay mineral fillings	6.0
H	Medium or low over-consolidation, softening, clay mineral fillings	8.0
J	Swelling-clay fillings, i.e., montmorillonite. Value of Ja depends on percent of swelling clay-size particles and access to water	8–12
(c)	No rock-wall contact when sheared (thick clay/crushed rock fillings	
M	Zones or bands of disintegrated or crushed rock and clay (see G, H, J for description of clay condition)	6, 8, or 8–12
N	Zones or bands of silty- or sandy-clay, small clay fraction (non-softening)	5.0
OPR	Thick, continuous zones or bands of clay (see G, H, J for description of clay condition)	10, 13, or 13–2

Table describes strength reduction factors (SRFc) formajor discontinuities such as faults, weakness zones, andjoint swarms which may also contain clay filling thatadversely affects slope stability. Major discontinuities mayor may not have a similar orientation to a discontinuity setsuch as a joint set or bedding plane. However, major dis-continuities are typically single features with considerablydifferent geomechanical properties (i.e., lower shearstrength due to soft, plastic in-filling).

Relationship Between Q-Slope and Slope Angles

Barton and Bar derived a simple formula for the steepest slope angle (b) not requiring reinforcement or support for slope heights less than 30 m. This formula is now extended to all slope heights:

$$\beta = 20\log_{10} Q_{slope} + 65\,°$$

Equation $\beta = 20\log_{10} Q_{slope} + 65\,°$ matches the central data for stable slope angles greater than 35°and less than 85°.

Table: Discontinuity orientation factor O-factor

O-factor description	Set A	Set B
Very favorably oriented	2.0	1.5
Quite favorable	1.0	1.0

Unfavorable	0.75	0.9
Very unfavorable	0.50	0.8
Causing failure if nsupported	0.25	0.5

It has a probability of failure of 1%. From the Q-slope data, the following correlations are simple and easy to remember:

Figure: Examples of the need for O-factor (discontinuity orientation factor) application

- Q-slope =10 -slope angle 85°

- Q-slope =1-slope angle 65°

- Q-slope =0.1 -slope angle 45°

- Q-slope =0.01 -slope angle 25°

Figure illustrates the available Q-slope data derived from the back-analysis of motorway, railway and road cuts, opencast mine benches, and natural slopes:

- Triangles indicate stable slopes. No visual signs of instability observed at least several weeks, months or years post-excavation.

Very favorably to favorably oriented joints forming cubical blocks: Norway.

Incipient, individually quite unfavorable relic joints forming wedge in weak saprolite: Panama.

Unfavorable foliation in phyllite. Steep joints striking into bench slopes are quite favorable: Western Australia.

Very unfavorable bedding planes in siltstone. Vertical joints and joints dipping into slope favorable. Mt. Isa.

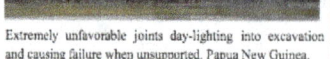

| Extremely unfavorable joints day-lighting into excavation and causing failure when unsupported. Papua New Guinea. | Extremely unfavorable joints day-lighting and causing failure when unsupported. Papua New Guinea. |

Table: Environmental and geological condition number

J_{wice}	Desertenvironment	Wetenvironment	Tropicalstorms	Ice wedging
Stable structure; competent rock	1.0	0.7	5	0.9
Stable structure; incompetent rock	0.7	0.6	0.3	0.5
Unstable structure; competent rock	0.8	0.5	0.1	0.3
Unstable structure; incompetent rock	0.5	0.3	0.05	0.2

When drainage measures are installed, apply J_{wice} X 1.5, when slope reinforcement measures are installed, apply J_{wice} 9 1.3, and when drainage and reinforcement are installed, apply both factors J_{wice} 9 1.5 X 1.3

Table: SRFa physical condition

Description		SRFa
A	Slight loosening due to surface location, disturbance from blasting or excavation	2.5
B	Loose blocks, signs of tension cracks and joint shearing, susceptibility to weathering, severe disturbance from blasting	5
C	As B, but strong susceptibility to weathering	10
D	Slope is in advanced stage of erosion and loosening due to periodic erosion by water and/or ice-wedging effects	15
E	Residual slope with significant transport of material downslope	20

Table: SRFb stress and strength

Description		σ_c/σ_1	SRFb
F	Moderate stress-strength range	50–200	2.5–1
G	High stress-strength range	10–50	5-2.5
H	Localized intact rock failure	5–10	10–5
J	Crushing or plastic yield	2.5–5	15–10
K	Plastic flow of strain softened material	1–2.5	20–15

σ_c = unconfined compressive strength (UCS), σ_1 = maximum principal stress

- Squares indicate quasi-stable slopes (more than likely to collapse soon with rainfall or weathering effects). Visible signs of slope instability such as tension cracks, dislocation, or deformation by means of monitoring (survey prisms or surface extensometers) are being continuously observed.

- Crosses indicate failed or collapsed slopes that have been back-analyzed using known pre-failure geometries and ground conditions.

Figure also includes larger slopes such as inter-ramp slopes in opencast mines where geological units and rock mass quality are uniform or very close to uniform across the height of the slope.

Equation $\beta = 20\log_{10} Q_{slope} + 65\,°$ does not represent a specific factor of safety as would be obtained by undertaking numerical analyses. Rather it represents the boundary of long-term stable slopes based on observed performance, normally between 6 months and over 50 years. However, users may, if they wish, additionally apply a factor of safety on the steepest slope angle (b) not requiring reinforcement or support. Considering only the failed and quasi-stable slopes, both of which are undesirable or unwanted events, the probability of failure (PoF) was calculated and is displayed using iso-potential lines in figure. The authors acknowledge that these iso-potential lines are one possible interpretation of the data and that other similar interpretations are also possible. If certain degrees of failure are accepted, such as percentages of individual benches in opencast mines, then the following equations can be derived:

SRF_c		Favorable	Unfavorable	Very unfavorable	Causing failure if unsupported
L	Major discontinuity with little or no clay	1	2	4	8
M	Major discontinuity with RQD100 = 0[a] due to clay and crushed rock	2	4	8	16
N	Major discontinuity with RQD300 = 0[b] due to clay and crushed rock	4	8	12	24

[a] RQD100 = 1 m perpendicular sample of discontinuity, [b] RQD300 = 3 m perpendicular sample of discontinuity

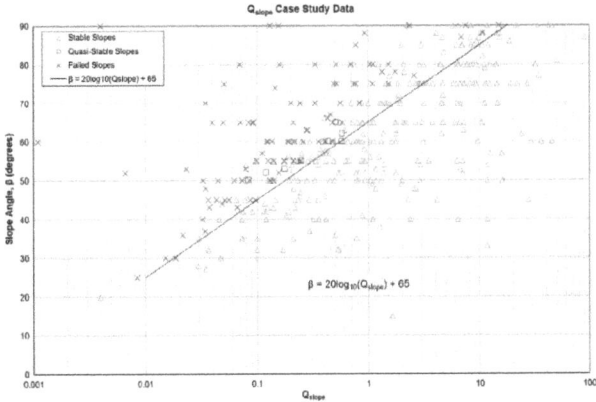

Figure: Q-slope data—412 case studies

$$PoF = 1\% : \beta = 20\log_{10} Q_{slope} + 65°$$

$$PoF = 15\% : \beta = 20\log_{10} Q_{slope} + 67.5°$$

$$PoF = 30\% : \beta = 20\log_{10} Q_{slope} + 70.5^\circ$$

$$PoF = 50\% : \beta = 20\log_{10} Q_{slope} + 73.5^\circ$$

Q-Slope Examples

Q-slope can be applied irrespective of rock strength, degree of fracturing, degree of weathering, etc. It also remains unchanged whether it is being used as a predictive or retrospective analysis. Q-slope cannot be applied to soil masses, rock fill, or landslide debris.

Planar Sliding

A 12 m high slope was excavated with a slope angle of 55° very strong quartzite ($\sigma_c > 150 MPa$). The outward dipping bedding ($\sim 50^\circ$) caused planar failure a few hours after excavation as illustrated in figure above. The following Qslope ratings were assigned during back-analysis:

- RQD = 90–100%

- J_n = 12

- J_r = 1, J_a = 3, O-factor = 0.25 (Set A only)

J_{wice} = 0.8 (desert environment, competent rock but unstable structure).

SRF_a = 2.5, SRF_b = 1, SRF_c = 1.

Based on the assigned ratings, Q-slope and b were estimated as follows:

$$Q_{slope} = \frac{95}{12} \times \left(\frac{1}{3} \times 0.25 \right) \times \frac{0.8}{2.5} = 0.211$$

$$\beta = 20\log_{10}(0.211) + 65^\circ = 51^\circ$$

Q-slope suggests an angle of 51° would have resulted in a stable slope. Given that the slope failed along a bedding plane dipping at approximately 50°, the back-analysis is considered sensible.

Wedge Sliding

A 30 m high slope was excavated at an angle of 65° and failed shortly after. The wedge failure occurred in weak, moderately weathered sandstone ($\sigma_c = 35 MP_a$) as illustrated in Figure. The following Q-slope ratings :

RQD = 40–50%

- J_n = 9

- Set A: J_r = 1, J_a = 4, O-factor = 0.5

- Set B: J_r = 3, J_a = 4, O-factor = 0.9

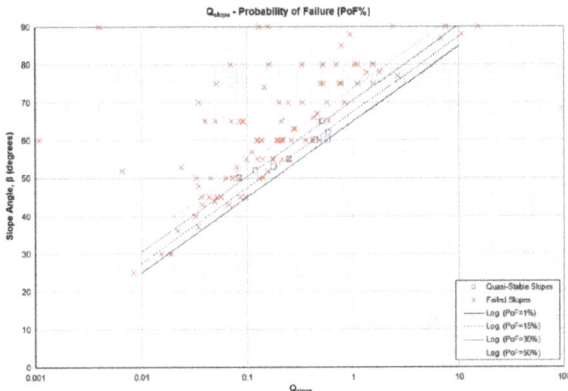

Figure: Q-slope probability of failure based on unwanted events
(failed or quasi-stable slopes are both undesirable)

Figure: Plane failure in quartzite bench (slope height = 12 m) from a large opencast mine in Western Australia. b Wedge failure in sandstone (slope height = 30 m). c Stable limestone slope in Serbia (slope height = 70 m

- Set C: Release plane or tension crack that did not contribute to the overall shear strength of the wedge.

- J_{wice} = 1 (desert environment, competent rock, and generally stable structure where Set B has limited continuity).

- SRF_a = 2.5 (slight loosening due to surface location),SRF_b = 2.5, SRF_c = N/A.

Based on the assigned ratings, Q-slope and β were estimated as follows:

$$Q_{slope} = \frac{55}{9} \times \left[\left(\frac{1}{4} \times 0.5 \right) \left(\frac{3}{4} \times 0.9 \right) \right] \times \frac{1}{2.5} = 0.206$$

$$\beta = 20 \log_{10}(0.206) + 65^{o} = 51^{o}$$

Q-slope suggests an angle of 51° would have resulted in a stable slope (i.e., approximately 51° shallower than excavated and consistent with kinematic analysis).

Steep Stable Slope

A 70 m high slope has been stable for at least 70 years following road construction at an angle of approximately 70° in strong limestone ($\sigma_c = 70 MP_a$) as illustrated in figure. The following Q-slope ratings were assigned during back-analysis:

- RQD = 90–100% (relatively uniform across entire slope measured at base and rest).

- J_n = 9.

- J_r = 3, J_a = 3, O-factor = 1 (Set A only).

- J_{wice} = 0.9 (alpine environment—ice wedging and freeze–thaw in winter, competent rock and stable structure).

- SRF_a = N/A, SRF_b = 3, SRF_c = 1 (minor shears, favorably oriented).

Based on the assigned ratings, Q-slope and β were estimated as follows:

$$Q_{Slope} = \frac{95}{9} \times \left(\frac{3}{3} \times 1.0 \right) \times \frac{0.9}{3} = 3.167.$$

$$\beta = 20 \log_{10}(3.167) + 65° = 75°$$

Q-slope suggests angles up to 75° would be stable.

Weak Rocks and Increasing Slope Heights

Residential subdivision earthworks cuttings in Far North Queensland often comprise weak weathered rocks and saprolites since excavations are usually relatively shallow. Figure is an example of such a slope excavated 5 m high at an angle of 35° without any form of geotechnical investigation or design. The slope comprised topsoil at the surface with the remainder being low strength saprolitic phyllite ($\sigma_c = 1 - 2 MPa$)). Relic geologic structure was clearly visible, even after excavation, and was quite favorably oriented. The following Q-slope ratings were assigned during back-analysis:

- RQD = 10% (minimum value)

- J_n = 6

- J_r = 2, Ja = 1, O-factor = 1 (Set A only)

- J_{wice} = 0.6 (wet environment—incompetent rock and stable structure).

- SRF_a = N/A, SRF_b = 7, SRF_c = N/A (slope height of 5 m results in SRF_b = 7).

Based on the assigned ratings, Q-slope and β were estimated as follows:

$$Q_{slope} = \frac{10}{6} = (\frac{2}{3} \times 1.0) \times \frac{0.6}{7} = 0.095.$$

$$\beta = 20\log_{10}(0.095) + 65° = 45°$$

Q-slope suggests angles up to $45°$ would be stable for a 5 m high slope. In this instance, the slope was steepened to $45°$ while retaining the crest position. Further proposed earthworks required the slope height to be increased to approximately 20 m. Since, the slope height has changed, Q-slope ratings as before were assigned, assuming ground conditions will remain similar as excavation depths increased:

SRF_a = N/A, SRF_b = 16, SRF_c = N/A (slope height of 20 m results in SRF_b = 16).

Based on the assigned ratings, Q-slope and β were estimated as follows:

$$Q_{slope} = \frac{10}{6} \times \left(\frac{2}{3} \times 1.0\right) \times \frac{0.6}{16} = 0.042.$$

$$\beta = 20\log_{10}(0.042) + 65° = 37°$$

Q-slope suggested the overall slope of 20 m in height could have a maximum angle of $37°$. This was supported by limit equilibrium analysis and a benched slope with an overall angle of $37°$.

Machine Foundations

Machine foundations are special foundations that are required for machine tools, machines and heavy equipment, which operate under varied loads, speeds and conditions. Such foundations are designed as per the dynamic forces that result from the operation of these machines. The topics elaborated in this chapter on vibration, machine foundation analysis and design, vibration analysis of machine foundation, design charts for machine foundation, damping, etc. will help in providing a comprehensive understanding of machine foundations.

The familiarity with the Foundations subjected to static loads has been very common. But, in some cases the foundations are also subjected to the dynamic loads. The resulting dynamic forms of loads may be from numerous causes like vibratory motion of machines, impacts from hammering, vehicle access/movement, earthquakes, waves, winds, cyclones, nuclear blasts, mining explosions, pile driving process and many more. The spread or transfer of these dynamic loadings/vibrations to the foundations and their consequences on the soil strata below can be determined using the principle of soil dynamics and theory of vibrations. The type of foundations used to encounter such dynamic forces/vibrations caused by machines used is known as "machine foundations". The dynamic forces are transmitted to the foundation supporting the machine. Generally the moving, shaking components of the machines are balanced, yet there is always some unbalanced condition in actual practice which results the eccentricity of rotating parts. Thus, it is very essential that the machine foundations require satisfying the criteria of dynamic loading plus static loading that exists already.

A machine foundation necessitates a special concern as they are designed to transmit dynamic loads to soil strata in addition to static loads. Event thought the dynamic loading/forces/vibrations due to machine operations is normally small in comparison to the static weight/loads from the machine and the supporting foundation. In machine foundations the dynamic loads are applied continually over a lengthier period of timing however since its magnitude is comparatively small so the soil behavior is found to be essentially elastic, otherwise deformations tends to increase after each cycle of loading and may lead to be incredibly higher that becomes unacceptable. On top of the natural frequency of the machine foundation soil system the amplitude of vibration of the machine during its working frequency is the most vital parameter to be considered and determined while designing of a machine foundation.

The general principles of machine design are enumerated as follows:

1. Machine foundations should be isolated from the adjoining parts of the building by leaving a gap around it to avoid the transmission of vibrations. The gap is filled with suitable insulator or dampers.

2. The foundation should be stiff and rigid to avoid possibilities of tilt in it.

3. In static state, the resultant of forces acting on machine foundation should pass through

the c.g of the contact area of the base.

4. The weight of the foundation block should be adequate. It should be able to absorb vibrations and resist resonance between adjoining soil and the machine. The weight of the foundation block may be assumed roughly 2.5 times the weight of the machine.

5. A vibration absorbing medium is introduced between bottom of the foundation block and the floor on which it is resting. This medium may be in form of a rubber or leather gasket, layer of sawdust etc. Sometimes, springs are used below the machine itself to prevent the development of the vibrations. It is advised that relevant data of machine should be obtained from the manufacturer before any design of foundation is undertaken.

Types of Machine

Rotating machinery: This category includes gas turbines, steam turbines, and other expanders; turbo-pumps and compressors; fans; motors; and centrifuges. These machines are characterized by the rotating motion of impellers or rotors. Unbalanced forces in rotating machines are created when the mass centroid of the rotating part does not coincide with the center of rotation. This dynamic force is a function of the shaft mass, speed of rotation, and the magnitude of the offset. The offset should be minor under manufactured conditions when the machine is well balanced, clean, and without wear or erosion. Changes in alignment, operation near resonance, blade loss, and other alfunctions or undesirable conditions can greatly increase the force applied to its bearings by the rotor. Because rotating machines normally trip and shut down at some vibration limit, a realistic Continuous dynamic load on the foundation is that resulting from vibration just below the trip level. Reciprocating machinery: For reciprocating machinery, such as compressors and diesel engines, a piston moving in a cylinder interacts with a fluid through the Kinematics of a slider crank mechanism driven by, or driving, a rotating crankshaft. Individual inertia forces from each cylinder and each throw are inherently unbalanced with dominant frequencies at one and two times the rotational frequency. Reciprocating machines with more than one piston require a particular crank arrangement to minimize unbalanced forces and moments. A mechanical design that satisfies operating requirements should govern. This leads to piston/ cylinder assemblies and crank arrangements that do not completely counter-oppose; therefore, unbalanced loads occur, which should be resisted by the foundation. Individual cylinder fluid forces act outward on the cylinder head and inward on the crankshaft. For a rigid cylinder and frame these forces internally balance, but deformations of large machines can cause a significant portion of the fluid load to be transmitted to the mounts and into the foundation. Particularly on large reciprocating compressors with horizontal cylinders, it is inappropriate and unconservative to assume the compressor frame and cylinder are sufficiently stiff to internally balance all forces. Such an assumption has led to many inadequate mounts for reciprocating machines.

Impulsive machinery: Equipment, such as forging hammers and some metal-forming presses, operate with regulated impacts or shocks between different parts of the equipment. This shock loading is often transmitted to the foundation system of the equipment and is a factor in the design of the foundation.

Closed die forging hammers typically operate by dropping a weight (ram) onto hot metal, forcing it into a predefined shape. While the intent is to use this impact energy to form and shape the material, there is significant energy transmission, particularly late in the forming process. During these final blows, the material being forged is cooling and less shaping takes place. Thus, pre-impact kinetic energy of the ram converts to post-impact kinetic energy of the entire forging hammer. As the entire hammer moves downward, it becomes a simple dynamic mass oscillating on its supporting medium. This system should be well damped so that the oscillations decay sufficiently before the next blow. Timing of the blows commonly range from 40 to 100 blows per min. The ram weights vary from a few hundred pounds to 35,000 lb (156 kN). Impact velocities in the range of 25 ft/s (7.6 m/s) are common. Open die hammers operate in a similar fashion but are often of two-piece construction with a separate hammer frame and anvil. Forging presses perform a similar manufacturing function as forging hammers but are commonly mechanically or hydraulically driven. These presses form the material at low velocities but with greater forces. The mechanical drive system generates horizontal dynamic forces that the engineer should consider in the design of the support system. Rocking stability of this construction is important. Figure shows a typical horizontal forcing function through one full stroke of a forging press.

Mechanical metal forming presses operate by squeezing and shearing metal between two dies. Because this equipment can vary greatly in size, weight, speed, and operation, the engineer should consider the appropriate type. Speeds can vary from 30 to 1800 strokes per min. Dynamic forces from the press develop from two sources: the mechanical balance of the moving parts in the equipment and the response of the press frame as the material is sheared (snap-through forces). Imbalances in the mechanics of the equipment can occur both horizontally and vertically. Generally high-speed equipment is well balanced. Low-speed equipment is often not balanced because the inertia forces at low speeds are small. The dynamic forces generated by all of these presses can be significant as they are transmitted into the foundation and propagated from there.

Other machine types: Other machinery generating dynamic loads include rock crushers and metal shredders. While part of the dynamic load from these types of equipment tend to be based on rotating imbalances, there is also a random character to the dynamic signal that varies with the particular operation.

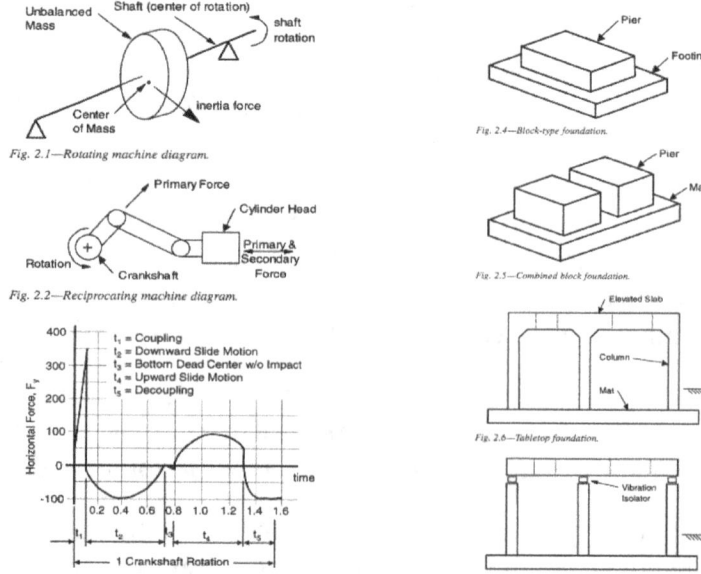

Figure: Forcing function for a forging press.   Figure: Table top with isolators.

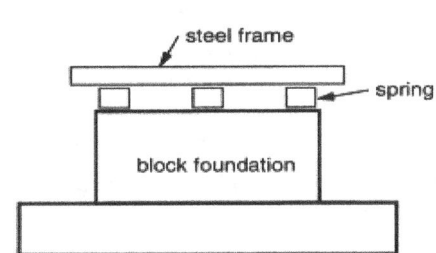

Figure: Spring-mounted block formation.

Types of Machine foundation

Block-type foundation: Dynamic machines are preferably located close to grade to minimize the elevation difference between the machine dynamic forces and the centre of gravity of the machine-foundation system. The ability to use such a foundation primarily depends on the quality of near surface soils. Block foundations are nearly always designed as rigid structures. The dynamic response of a rigid block foundation depends only on the dynamic load, foundation's mass, dimensions, and soil characteristics.

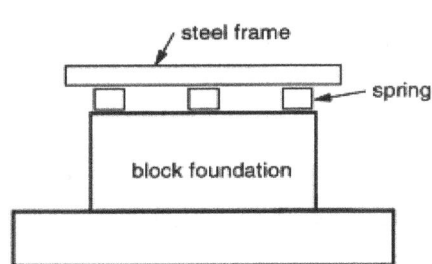

Figure: Spring-mounted block formation.

Combined block-type foundation: Combined blocks are used to support closely spaced machines. Combined blocks are more difficult to design because of the combination of forces from two or more machines and because of a possible lack of stiffness of a larger foundation mat.

Tabletop-type foundation: Elevated support is common for large turbine-driven equipment such as electric generators. Elevation allows for ducts, piping, and ancillary items to be located below the equipment. Tabletop structures are considered to be flexible, hence their response to dynamic loads can be quite complex and depend both on the motion of its discreet elements (columns, beams, and footing) and the soil upon which it is supported. Tabletop with isolators: Isolators (springs and dampers) located at the top of supporting columns are sometimes used to minimize the response to dynamic loading. The effectiveness of isolators depends on the machine speed and the natural frequency of the foundation.

Spring-mounted equipment: Occasionally pumps are mounted on springs to minimize thermal forces from connecting piping. The springs are then supported on a block-type foundation. This arrangement has a dynamic effect similar to that for tabletops with vibration isolators. Other types of equipment are spring mounted to limit the transmission of dynamic forces.

Inertia block in structure: Dynamic equipment on a structure may be relatively small in comparison to the overall size of the structure. In this situation, dynamic machines are usually designed with a supporting inertia block to alter natural frequencies away from machine operating speeds and resist amplitudes by increasing the resisting inertia force.

Pile foundations: Any of the previously mentioned foundation types may be supported directly on soil or on piles. Piles are generally used where soft ground conditions result in low allowable contact pressures and excessive settlement for a mat-type foundation. Piles use end bearing, frictional side adhesion, or a combination of both to transfer axial loads into the underlying soil. Transverse loads are resisted by soil pressure bearing against the side of the pile cap or against the side of the piles. Various types of piles are used including drilled piers, auger cast piles, and driven piles.

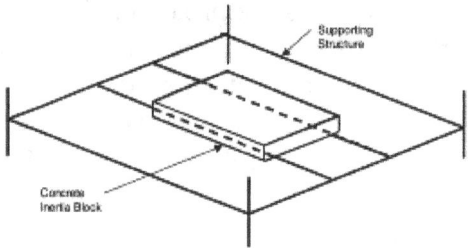

Figure: Inertia blocks in structure.

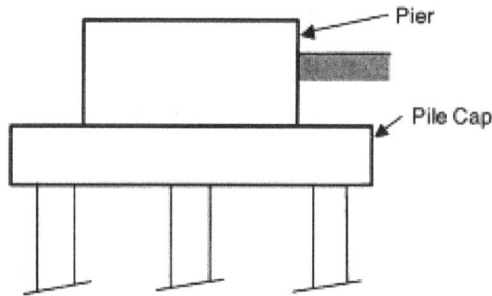

Figure: Pile-supported foundation.

Vibration

Certain fundamental aspects of Vibration essential to the study of Soil Dynamics are considered in the following subsections.

Degree of Freedom

The 'Degree of Freedom' for a system is defined as the minimum number of independent co-ordinates required to describe the motion of the system mathematically. A mass supported by a spring and constrained to move in only one direction is a system with a single degree of freedom. Similarly, a simple pendulum oscillating in one plane is also an example of a system with a single degree of freedom.

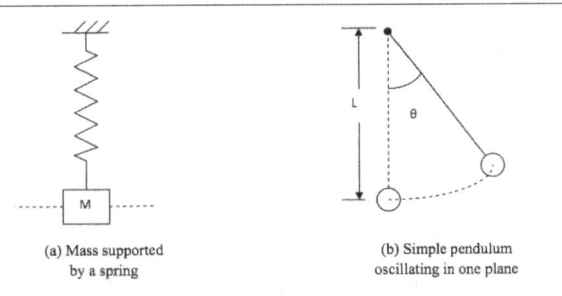

(a) Mass supported
by a spring

(b) Simple pendulum
oscillating in one plane

Figure: Systems with single degree of freedom

However, if the spring-supported mass of figure can also rotate in one plane its degree of freedom is two. A two-mass two-spring system, constrained to move in one direction without rotation, is also an example of a system with a degree of freedom of two. A body in space has a degree of freedom of six—three translational and three rotational. A flexible beam between two supports has an infinite number of degrees of freedom.

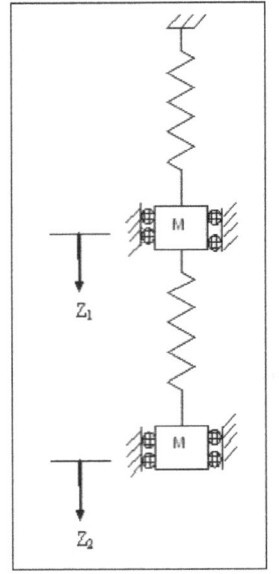

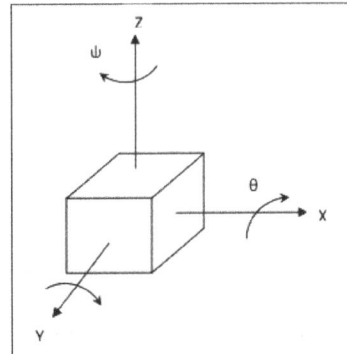

Figure: Body in space with six degrees of freedom

Figure: A two-mass two-spring mass system

Figure: A beam with infinite degree of freedom

Modes of Vibration

A system with more than one degree of freedom vibrates in complex modes. However, if each point in the system follows a definite pattern of vibration, the mode is systematic and orderly, and is known as a 'principal mode of vibration'. The vibration of a block can be reduced to six modes for the purpose of analysis. These are:

   (i) Translation along X-axis (lateral)

  (ii) Translation along Y-axis (longitudinal)

 (iii) Translation along Z-axis (vertical)

(iv) Rotation about X-axis (pitching)

(v) Rotation about Y-axis (rocking)

(vi) Rotation about Z-axis (yawing or torsional)

These are known as the Principal modes of vibration of the block and are shown Schematically in figure can occur independently but not others. This is because rotation about X-axis or Y-axis is always accompanied by translation along Y- or X-axis and viceversa, producing what is known as 'coupled motion'. If a combination of more than one mode of vibration occurs in a particular case, it is referred to as 'coupled mode' of vibration. The analysis of such modes requires the use of complex mathematical treatment.

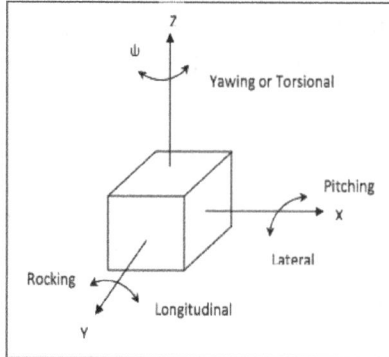

Figure: Modes of vibration of a block

Free Vibrations and Forced Vibrations

Bodies which have both mass and elasticity are capable of undergoing vibrations. The vibrations of a body or a system may be classified as 'Free Vibrations' and 'Forced Vibrations'. 'Free Vibration' is a vibration that occurs under the influence of forces inherent in the system itself, without any external force. Of course, an external force or natural disturbance is required to initiate the free vibration which continues without an external force acting continuously. If the vibration is un-damped by friction or any other forces, the body undergoes free vibration with a frequency known as the 'Natural frequency' of the body or system. It is considered as the property of the body or system. Depending upon the particular mode of vibration, the body will have a particular value of natural frequency. Thus a body or system can have as many natural frequencies as the possible modes of vibration. 'Forced Vibration' is a vibration that occurs under the continuous influence of an external force. This obviously depends upon the nature of the external force, also known as the 'exciting force', which may be caused by an impulsive force or a continuous periodic force. Hammer foundation produces an impulsive force causing forced vibration of the system. A foundation for a machine with rotating masses will be subjected to a vibration caused by a continuous periodic force. In practice it is extremely rare that a body has free vibration at its natural frequency undamped, since it is always subjected to some form of damping.

Resonance

When the frequency of the exciting force in a forced vibration of a body or a system equals one of the natural frequencies of the body or system, the amplitude of motion tends to become excessively large. This condition or phenomenon is called 'Resonance'. The particular value of the frequency of the exciting force producing resonant conditions is called the 'resonant frequency' under that specific mode of vibration. Since resonance produces excessively large amplitudes, it has dangerous implications for any engineering structure, machine, or system in causing failure. Hence one of the important endeavours of an Engineer dealing with Soil Dynamics and Design of a Machine Foundation is to avoid resonant conditions.

Damping

'Damping' in a physical system is resistance to motion, and may be one of the several types mentioned in the following paragraphs. Viscous Damping: - This type of damping occurs in lubricated sliding surfaces, dashpots with small clearances etc. Eddy current damping is also of viscous nature. The magnitude of damping depends upon the relative velocity and upon the parameters of the damping system. For a particular system, the damping resistance is proportional to the velocity:

$$F = c\frac{dz}{dt}$$

where, F =damping force $\frac{dz}{dt}$ =velocity, and c=damping coefficient.

a) This affords relatively easy analysis of the system, since the differential equation of the system becomes linear with this type of damping. This is why a system is often represented to include an equivalent viscous damper even if the damping is not truly viscous.

Friction or Coulomb Damping: This kind of damping occurs when two machine parts rub against each other, dry or un-lubricated. The damping force in this case is practically constant and is independent of the velocity with which the parts rub each other.

b) Solid, Internal or Structural Damping: This type of damping is due to the internal friction of the molecules. The stress-strain diagram for a vibrating body is not a straight line but forms a hysteresis loop, the area of which represents the energy dissipated due to molecular friction per cycle per unit volume. The area of the loop depends upon the material of the vibrating body, frequency, and the magnitude of the stress. Since this involves internal loss of energy by absorption, it is also called 'internal damping'.

c) Slip or Interfacial Damping: Energy of vibration is dissipated by microscopic slip on the interfaces of machine parts in contact under fluctuating loads. Microscopic slip also occurs on the interfaces of the machine elements forming various types of joints. The magnitude of damping depends, amongst other things, upon the surface roughness of the parts, the contact pressure, and the amplitude of vibration. This type of damping is essentially of a non-linear type.

d) Radiation', 'dispersion' or 'geometric' damping: In the case of machine foundation resting on soil, damping occurs due to the loss of energy on two counts. First, some energy loss occurs by the absorption of energy into the system, reflected by the hysteresis in the stress strain relationship; damping caused by this internal loss of energy is called 'internal damping' already given in (b). Next, the dissipation of energy by wave propagation radiating away into the soil mass, causes damping effect. This is known as 'radiation', 'dispersion', or 'geometric' damping.

Negative Damping

Generally speaking, damping is positive, so that energy is always absorbed from the system by damping devices. If the system draws energy from some source or is supplied energy, the amplitude continues to increase, leading to instability. Such a system is said to be negatively damped. The build-up of amplitudes of transmission line wires, or tall poles or suspension bridges under the action of uniform wind flow at critical speeds are examples of negatively damped systems. In structural systems subjected to dynamic forces due to an earthquake or a blast, the damping is always positive.

Free Vibration with Damping

The mathematical model consists of a mass supported by a weightless spring with single degree freedom.

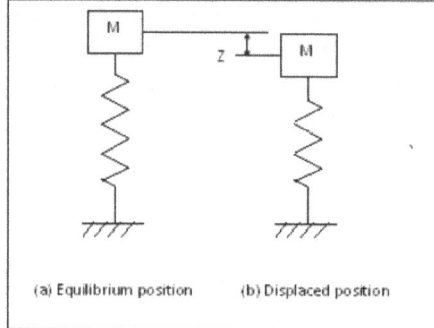

Figure: Free vibrations- undamped-mass spring system

If z is the vertical displacement of the system from its equilibrium position, and k is the spring constant, applying Newton's law of motion, the equation of motion is:

$$M\ddot{Z} + KZ = 0$$

Or,

$$\ddot{Z} + \left(\frac{k}{M}\right) z = 0$$

Or

$$\ddot{Z} + \omega_n^2 Z = 0$$

Where

$$\omega_n^2 = \frac{k}{M}$$

Equation($\ddot{Z} + \omega^2{}_n = 0$) is a homogeneous linear differential equation and the solution is given by;

$$z = c_1 \sin \omega t + C_2 \cos \omega t$$

where, C_1 and C_2 are constants and can be evaluated from the initial conditions of the system. The equation also represents simple harmonic motion expressed by $z = c_1 \sin \omega t + C_2 \cos \omega t$, ωn being the circular frequency. Therefore, the free vibration of a mass resting on a spring and subjected to inertial forces only can be represented by a simple harmonic motion.

ω_nb in this case is called 'Natural Circular Frequency' of the system.

$$\omega_n = \sqrt{\frac{k}{M}} rad / sec$$

And

$$f_{n=}\frac{\omega_n}{2\pi} = \frac{1}{2\pi}\sqrt{\frac{k}{M}}$$

The Period

$$T_{n=}\frac{2\pi}{w_n} = 2\pi\sqrt{\frac{M}{k}}$$

Forced Vibration without Damping

If a mass supported by a spring is subjected to an exciting force, the system undergoes forced vibrations. Such an exciting force may be caused by unbalanced rotating machinery or by other means. In the analysis that follows, it is assumed that the exciting force is periodic and that it may be expressed as

$$P=P_0 \, sin \, \omega t$$

where, P_0 is the maximum value of the exciting force and ω is the circular frequency of the exciting force in rad/sec. The system is shown in figure below.

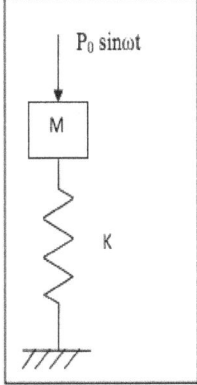

Figure: Forced Vibration- undamped mass-spring system

The equation of motion for the system may be written as;

$$M\ddot{Z} + kZ = P_0 \, sin \, \omega t$$

Or,

$$\ddot{Z} + w^2_{\,n}z = \frac{P_0}{M} sin \, \omega t$$

Since

$$w_n^2 = \frac{k}{M}$$

The solution of equation above includes the solution for free vibrations, along with the solution which satisfies the right hand side of Eq.($M\ddot{Z} + kZ = P_o \sin \omega t$). The solution may be obtained by parts as the sum of the complementary function and the particular integral. The complementary function which represents the free vibration does not exist in this situation and the particular integral alone is of interest. Since the applied force is harmonic, the motion of the system may be taken as being harmonic. Thus the particular integral may be taken as

$$Z = A \sin \omega t$$

By substituting this in $M\ddot{Z} + kZ = P_o \sin \omega t$, we may show that

$$A = \frac{P_o}{M\left(\omega_n^2 - \omega^2\right)}$$

It follows that the frequency of a forced vibration is equal to that of the exciting force. (This is the same as the speed of machine, in case it is a machine that is being dealt with). equation above may be rewritten as;

$$A = \frac{P_o}{M\omega_n^2\left(1 - \frac{\omega^2}{\omega_n^2}\right)}$$

But

$$\frac{P_o}{M\omega_n^2} = A_{st}$$

where, A_{st} =deflection of the system under P_o, applied statically. The ratio $\left(\frac{\omega}{\omega_n}\right)$ is called the frequency ratio, ξ .

$$A = \frac{A_{st}}{1 - \xi}$$

The factor $\left(\frac{1}{1-\xi}\right)$ is called the 'magnification factor', η_o is shown in figure above.

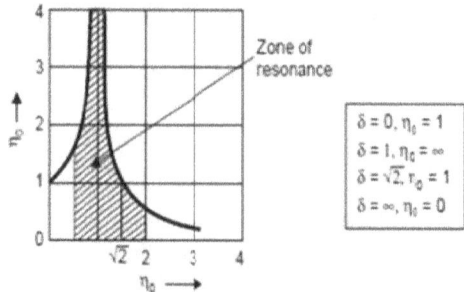

Figure: Frequency ratio VS magnification factor

When the exciting frequency approaches the natural frequency of the system the magnification factor $(\xi = 1)$ and hence the amplitude of vibration tend to become infinite, leading to resonance. If the frequency ratio is more than 1, there will be steep decrease of the magnification factor.

It is obvious that resonant conditions should be avoided.

Free Vibrations with Damping

Assuming that in a system undergoing free vibrations viscous damping is present, a "mass-spring dashpot" system can serve as the relevant mathematical model for analysis. The 'dashpot' is the simplest mathematical element to simulate a viscous damper. The force in the dashpot under dynamic loading is directly proportional to the velocity of the oscillating mass.

The equation of motion is

$$M\ddot{Z} + c\dot{Z} + kZ = 0$$

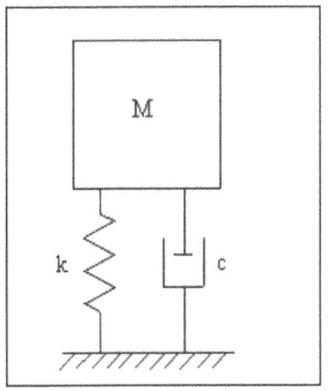

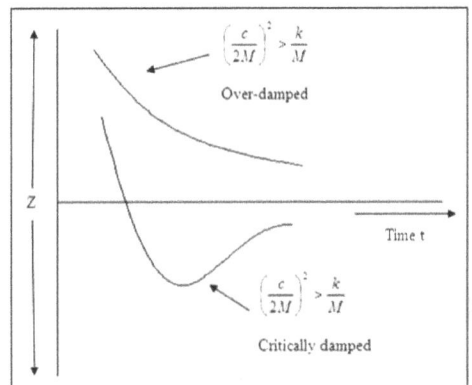

(a)Mass-spring dash pot system

(b)Different damping conditions over-damped and critically damped systems

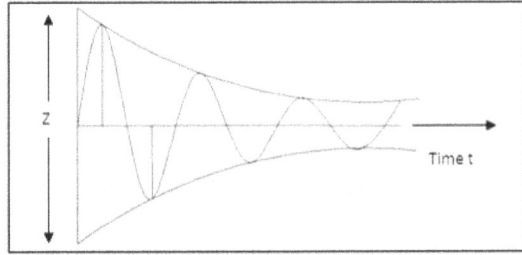

(c) Undamped system

Figure: A mathematical model for free vibrations with damping

This can be rewritten as

$$\ddot{Z} + \frac{c}{M}\dot{z} + \frac{k}{M}z = 0$$

Putting

$$\alpha = \frac{c}{M}$$

$$\ddot{Z} + \alpha \dot{Z} + \omega^2_n Z = 0$$

Let the solution to equation above be in the form

$$Z = e^{\lambda t}$$

λ being a constant to be determined.

Substituting this in Eq. ($\ddot{Z} + \alpha \dot{Z} + \omega^2_n Z = 0$) we get

$$\left(\lambda^2 + a\lambda + \omega^2_n\right)e^{\lambda t} = 0$$

Or, $\lambda^2 + a\lambda + \omega^2_n = 0$

The roots of this equation are

$$\lambda 1 = -\frac{a}{2} + \sqrt{\left(\frac{a}{2}\right)^2 - \omega_n^2}$$

$$\lambda 2 = -\frac{a}{2} + \sqrt{\left(\frac{a}{2}\right)^2 - \omega_n^2}$$

Three possible case of damping arise from these roots, these are:

Case-1: Roots are real and negative if

$$\left(\frac{\alpha}{2}\right)^2 = \omega^2_n \, or \left(\frac{c}{2M}\right)^2 = \frac{k}{M}$$

The general solution is

$$Z = C_1 e^{(\lambda_1 t)} + C_2 e^{(\lambda_2 t)}$$

Since both λ_1 and λ_2 are negative, z will decrease exponentially with time without any change in sign as shown in figure below. The motion is not periodic and the system is said to be over-damped.

Case-2: Roots are equal if

$$\left(\frac{\alpha}{2}\right)^2 = \omega^2_n \, or \left(\frac{c}{2M}\right)^2 = \frac{k}{M}$$

The general solution is

$$Z = e^{\left[-\left(\frac{c}{2M}\right)t\right]}(C_1 + C_2 t)$$

This is similar to the over-damped case except that it is possible for the sign to change once as shown in undamped figure. This is also not a periodic motion and with increase in time, approaches zero. The value of 'c' for this condition is called the 'critical damping coefficient 'c_c'.

$$\text{Since } \left(\frac{c_c}{2M}\right)^2 = \frac{k}{M}$$

$$c_c = 2\sqrt{kM}$$

Using Eq. ($\omega_n = \sqrt{\dfrac{k}{M}} rad/sec$) we may write

$$c_c = 2M\omega_n$$

c_c is the limiting value for the motion to be periodic.

Case-3: Roots are complex conjugates if

$$\left(\frac{\alpha}{2}\right)^2 < \omega^2{}_n \, or \left(\frac{c}{2M}\right)^2 < \frac{k}{M}$$

By using Equation ($c_c = 2\sqrt{kM}$) the roots λ_1 & λ_2 become

$$\lambda_1 = \omega_n\left(-D+i\sqrt{1-D^2}\right)$$

$$\lambda_2 = \omega_n\left(-D+i\sqrt{1-D^2}\right)$$

Where $D = \dfrac{c}{c_c}$ and is called "Damping Ratio" or "Damping factor". Substituting these into Eq. ($Z = C_1 e^{(\lambda_1 t)} + C_2 e^{(\lambda_2 t)}$) and simplifying, the general solution becomes

$$Z = e^{-\omega_n Dt}(C_3 \sin \omega_n t\sqrt{1-D^2} + C_4 \cos \omega t\sqrt{1-D^2})$$

where C_3 and C_4 are arbitary constants

Equation above indicates that the motion is periodic and the decay in amplitude will be proportional to $e^{-\omega_n Dt}$ as shown by the dashed curve in figure of undamped system. Further Eq. $Z = e^{-\omega_n Dt}(C_3 \sin \omega_n t\sqrt{1-D^2} + C_4 \cos \omega t\sqrt{1-D^2})$ indicates that the frequency of free vibrations with damping is less than the natural frequency for undamped free vibrations, and that as D → 1, the frequency approaches zero. The relation between these two frequencies is given by

$$\omega_{dn} = \omega_n \sqrt{1-D^2}$$

where, ω = frequency of free vibration with damping. Figure above shows that there is a decrement in the successive peak amplitudes. Using Eq., ratios of successive peak amplitudes may be found.

Let Z_1 and Z_2 be the amplitudes of successive peaks at times t_1 & t_2 respectively as shown in figure above.

$$\frac{Z1}{Z2} = exp\left(\frac{2\pi D}{\sqrt{1-D^2}}\right)$$

'Logarithmic Decrement' is defined as

$$\delta = In\frac{Z_1}{Z_2} = \frac{2\pi D}{\sqrt{1-D^2}}$$

In words, logarithmic decrement is defined as the natural logarithm of the ratio of any two successive amplitudes of same sign in the decay curve obtained in free vibration with damping.

δ is approximately $2\pi D$, when D is small. Eq. above also indicates that, in viscous damping, the ratio of amplitudes of any successive peaks is a constant. It follows that the logarithmic decrement may be obtained from any two peak amplitudes Z_1 and n Z_{1+n} from the equation.

$$\delta = \delta = \frac{1}{2}In\frac{Z1}{Z_{1+N}}$$

Forced Vibration with Damping

A system which undergoes forced vibrations, and in which viscous damping is present, may be analysed by the mass-spring-dashpot model shown in figure below. The equation of motion for this system may be written as follows:-

$$M\ddot{Z} + c\dot{Z} + kZ = P_0 \sin\omega t$$

This may be written as

$$\ddot{Z} + \frac{c}{M}\dot{Z} + \frac{k}{M}Z = \frac{P_0}{M}\sin\omega t$$

Or, $\ddot{Z} + \alpha\dot{Z} + \omega^2_n Z = \frac{P_0}{M}\sin\omega t$

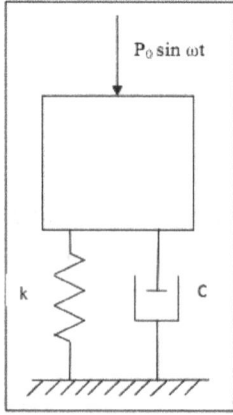

Figure: Forced vibration with damping

Where, $\alpha = \frac{c}{M}$ & $w^2_n = \frac{k}{M}$

The particular solution is a steady state harmonic oscillation having a frequency equal to that of the excitation, and the displacement vector lags the force vector by some angle. Let us therefore assume that the particular solution is:

$$Z = A\sin(\omega t - \phi)$$

Where $A = \dfrac{P_0}{M\omega_n^2 \sqrt{a^2 w^2 + (w_n^2 - w^2)}}$

and, $\phi = tan^{-1}\left[\dfrac{a\omega}{(w_n^2 - \omega^2)}\right]$

'A' may also be expressed as

$$A = \dfrac{P_0}{M\omega_n^2 \sqrt{(1-\xi^2) + 4D^2\xi^2}}$$

$\xi = \dfrac{\omega}{\omega_n}$ the frequency ratio

There are two kinds of excitation:

    i) Constant force-amplitude excitation and

    ii) Quadratic excitation.

Constant Force—Amplitude Excitation

This type is caused by an electro-magnetic vibrator, the exciting force being generated mainly from the magnetic attraction or repulsion due to the change in intensity or direction of magnetic flux linking several flux-carrying elements. The magnetic flux is produced by passing an electric current through a winding on one part of the magnetic circuit. The resultant magneto-motive force is proportional to the current passing through the coil. The other winding is placed in order to generate a force having fundamental frequency of the magnetic circuit and to eliminate the rectification process. An electromagnetic vibrator is driven by a frequency oscillator and power amplifier.

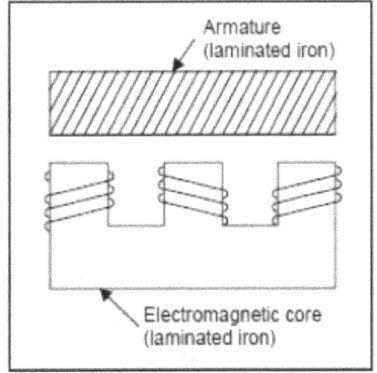

Figure: Electromagnetic Vibrator

Eq. ($A = \dfrac{P_0}{M\omega_n^2 \sqrt{(1-\xi^2)+4D^2\xi^2}}$) may be rewritten as follows:

$$A = \eta_1 A_{st}$$

where, $\eta = \dfrac{1}{\sqrt{(1-\xi)^2 + 4D^2\xi^2}}$

Since A is a constant for given spring and excitation, amplitude of motion A is directly proportional to η_1.

To determine the conditions corresponding to maximum amplitude, Eq. ($\eta = \dfrac{1}{\sqrt{(1-\xi)^2 + 4D^2\xi^2}}$) may be differentiated with respect to ξ, equated to zero, and solved for ξ. One obtains

$$\xi = \left(\sqrt{1-2D^2}\right)$$

It is clear from this equation that if D decreases, ξ increases and vice versa. Resonance condition is said to occur when the peak amplitude occurs. Hence, the magnification factor at resonance η_{1max} is got by substituting the value of ξ from Eq.($\xi = \left(\sqrt{1-2D^2}\right)$) in Eq.($\eta = \dfrac{1}{\sqrt{(1-\xi)^2 + 4D^2\xi^2}}$).

$$\eta_{1max} = \dfrac{1}{2D\sqrt{1-D^2}}$$

From this equation, it can be seen that the larger the damping ratio, the smaller the magnification factor at resonance, and vice versa. The relationship between ξ and η_1 (or A) [for varying D] is shown in figure below.

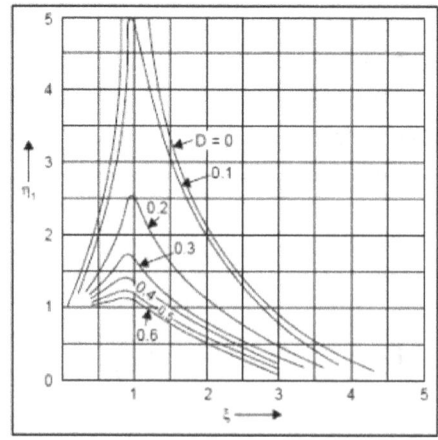

Figure: Magnification factor vs. frequency ratio

It can be observed that the maximum value of η_1, and hence the peak amplitude occurs at a value of ξ less than unity when damping is present. As the Damping ratio, D increases the value of ξ for peak amplitude deviates more from unity. The corresponding frequency at which peak amplitude occurs at a certain value of damping is known as the resonant frequency for the damped case.

It may be recalled that, without damping, the peak amplitude which occurs when $\xi = 1$, is infinite. The effect of damping is to make the peak amplitude finite and make the frequency ratio for peak amplitude deviate from unity. In other words, what is called the resonant frequency is different in the undamped and damped cases.

Quadratic Excitation

In this type of excitation, the exciting force is proportional to the square of the frequency. This is caused by the rotation of unbalanced masses (figure) in an oscillator.

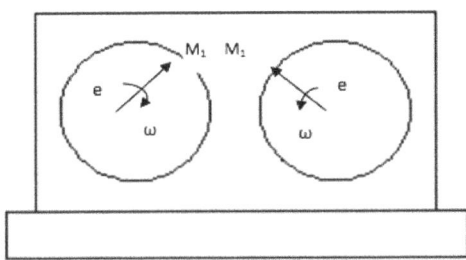
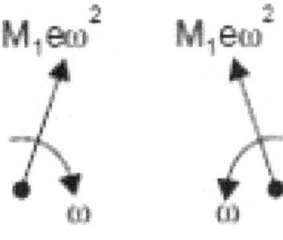

(a) Rotation of unbalanced masses   (b) Counteracting forces

Figure: Quadratic excitation due to rotation of unbalanced mass

The exciting moment, M_e, may be varied by varying either the total unbalanced mass M_e or the eccentricity e. The periodic force is not constant unlike the previous case. The rotating force of each mass is $M_1 e\omega^2$. The total force in the vertical position is $2M_1 e\omega$ or $M_e e\omega^2$ where M_e is the total unbalanced mass (equal to $2\,M_1$). The vibrating force at any position may be represented by

$$P = M_e e\omega^2 \sin\omega t = \bar{P}_0 \sin\omega t$$

$$\bar{P} = M_e e\omega^2 \text{ (Eq.)}$$

The periodic force is expressed by Eq. ($P = P_0 \sin\omega t$) replacing P_0 by $\bar{P}_0$ for frequency-dependent exciting force;

$$P = \bar{P}_0 \sin\omega t$$

We may write

$$\frac{\bar{P}_0}{k} = \frac{M_e e\omega^2}{k} = \left(\frac{M_e e}{M}\right)\left(\frac{M}{k}\right)w^2 = \frac{M_e e}{M}\left(\frac{\omega}{\omega_n}\right)^2 = \frac{M_e e}{M}\xi^2$$

The differential equation of motion and its solution are the same as those in the previous case in as much as these are independent of the method of applying the exciting force. The amplitude may be got as follows, using Eq. ($A = \dfrac{P_0}{M\omega_n^2\sqrt{(1-\xi^2)+4D^2\xi^2}}$), substituting P_0 for $\bar{P}$, and using Eq.

($\frac{\bar{P}_0}{k} = \frac{M_e e\omega^2}{k} = \left(\frac{M_e e}{M}\right)\left(\frac{M}{k}\right)w^2 = \frac{M_e e}{M}\left(\frac{\omega}{\omega_n}\right)^2 = \frac{M_e e}{M}\xi^2$), and simplifying further:

$$A = \frac{M_e e}{M} = \frac{\xi^2}{\sqrt{\left(1-\xi^2\right)^2 + 4D^2\xi^2}}$$

Analyzing in the same manner as in the previous case, the maximum amplitude occurs when

$$\xi = \frac{1}{\sqrt{1-2D^2}}$$

From this, it can be seen that as D decreases, ξ decreases and vice-versa. Defining magnification factor η_2 as;

$$\eta_2 = A. \frac{M}{M_e e} = \frac{\xi^2}{\sqrt{\left(1-\xi^2\right)^2 + 4D^2\xi^2}}$$

That means $\quad \eta_2 = \eta_1 \xi^2$

It can also be shown that, $\eta_{2max} = \dfrac{1}{2D\sqrt{1-D^2}}$

The relationship between ξ & η2 (or A) for different values of D is shown in figure.

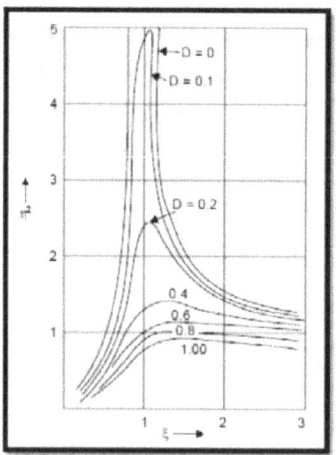

Figure: η2 versus ξ

It can be seen from this figure that for quadratic excitation the maximum value of η2 (or A)

occurs at a value of ξ greater than unity when damping is present. As the value of D increases, the peak η2 (or A) deviates more from ξ =1. Thus resonant conditions tend to occur at a frequency ratio more than unity.

In this case also, the effect of damping is to make the peak amplitude finite and to make

the ξ-value corresponding to the peak amplitude deviate from unity. It can also be seen that the influence of damping is large when resonance condition occurs and it decreases when the amplitude of motion is different from the peak amplitude; the greater the difference the smaller the influence of damping ratio.

Vibration Analysis of Machine Foundation

To analyze the vibration theory of machine foundation we need to assume that the machine foundation has single degree of freedom. Normally machine foundation has 6 degree of freedom.

Let us say a machine foundation is rest on soil mass. Now the mass of machine and foundation acts downwards together and it is say mf which acts at the center of gravity of the system. The mass of soil which acts upwards is say ms. the elastic action of soil due to vibration of system is dependent of stiffness k. Resistance against motion is dependent of damping coefficient c.

So, these three mass, stiffness and damping coefficient are required to complete the analysis of machine foundation. Determination of above parameters is explained below.

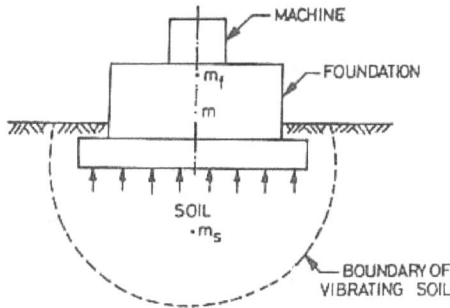

Mass (m)

Whenever the machine vibrates, soil below the machine foundation also vibrates. The mass of soil which vibrates due to machine vibration is termed as in-phase soil mass. Therefore, total mass (m) is equal to

$$m = m_f + m_s$$

Where,

m_f = mass of the foundation

m_s = in-phase soil mass = it varies from 0 to m_f

Total mass (m) varies from mf to 2mf.

Stiffness (k)

The stiffness is dependent of type of soil below the foundation, embedment of foundation block and contact pressure distribution between soil and foundation. Stiffness is derived from the following methods.

Laboratory Method of Vibration Analysis

In the laboratory, a tri axial test with vertical vibrations is performed and modulus of rigidity is obtained. From this young's modulus is determined with the help of Poisson's ratio.

Young's modulus E = 2G(1+u)

Stiffness k = AE/L

Where E = young's modulus

G = rigidity modulus

U = positions ratio

Barkan's Method

The stiffness can also be derived from the formula proposed by Barkan which is given below.

$$k = \frac{1.13E}{1-u2}\sqrt{A}$$

Where A = area of contact

Plate Load Test

A plate load test is conducted in the field and determine the stiffness of soil as slope of the load-deformation curve.

For cohesive soils, Stiffness K

$$k = k_p\left(\frac{B}{B_p}\right)$$

For cohesionless soils, Stiffness k

$$k = k_p\left(\frac{B+0.3}{B_p+0.3}\right)^2$$

Where,

B = width of foundation

Bp = diameter of plate

Resonance Test

By knowing the resonance frequency (f_n), we can calculate the stiffness value. f_n can be determined by placing vibrator of mass m on a steel plate supported on ground.

$$f_n = \frac{1}{2\pi}\sqrt{\frac{k}{m}}$$

Therefore, stiffness

$$k = 4\pi^2 f_n m.$$

General Requirements of Machine Foundations and Design Criteria

The following general requirements of machine foundations shall be satisfied and results checked prior to detailing the foundations:

1. The foundation should be able to carry the superimposed loads without causing shear or crushing failure.

2. The settlements should be within the permissible limits.

3. The combined centre of gravity of machine and foundation should, as far as possible, be in the same vertical line as the centre of gravity of the base plane.

4. No resonance should occur; hence the natural frequency of the foundation–soil system should be either too large or too small compared to the operating frequency of the machine. For low-speed machines, the natural frequency should be high.

5. The amplitudes under service conditions should be within permissible limits which are prescribed by the machine manufacturers.

6. All rotating and reciprocating parts of a machine should be so well balanced as to minimize the unbalanced forces or moments.

7. Where possible, the foundation should be planned in such a manner as to permit a subsequent alteration of natural frequency by changing base area or mass of the foundation as may subsequently be required.

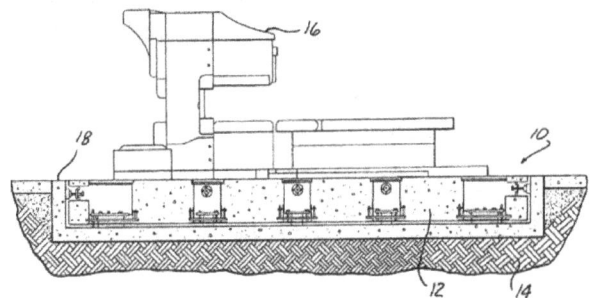

Figure: Machine Foundation

From the practical point of view, the following requirements should be fulfilled.

1. The groundwater table should be as low as possible and groundwater level deeper by at least one-fourth of the width of foundation below the base plane. This limits the vibration propagation, groundwater being a good conductor of vibration waves.

2. Machine foundations should be separated from adjacent building components by means of expansion joints.

3. Any steam or hot air pipes, embedded in the foundation must be properly isolated.

4. The foundation must be protected from machine oil by means of acid-resisting coating or suitable chemical treatment.

5. Machine foundations should be taken to a level lower than the level of the foundations of adjoining buildings.

Design Charts for Machine Foundation

The design of machine foundations is a trial-and-error procedure involving three interrelated steps

Gazetas and Roesset:

1) Establishment of desired foundation performance (design criteria),

2) Determination of magnitude and characteristics of the dynamic loading,

3) Estimation of anticipated translational and rotational motion of machine-foundation-soil system.

The design of a machine foundation is more complex than that of a foundation which supports only static loads. In machine foundations, the designer must consider, in addition to the static loads, the dynamic forces caused by the working of the machine operation. These dynamic forces are, in turn, transmitted to the foundation supporting the machine.

Design Limits of Machine Foundation for Empirical Methods

The design of block foundation for centrifugal or reciprocating machine starts with preliminary sizing of the block, which has been found to result in acceptable configuration as:

1. The bottom of block foundation should be above water table. It should not be resting on back filled soil nor on a special sensitive soil.

2. The mass of rigid foundation equals (2-3) times the mass of supported machine (for centrifugal), while the mass of rigid foundation equals (3-5) times the mass of supported machine (for reciprocating).

3. The top of block is usually kept (0.3 m) above finished floor or pavement elevation to prevent damage from surface water runoff.

4. The vertical thickness of block should not be less than (0.61 m). The thickness seldom less than one-fifth the least dimension or one-tenth the largest dimension.

5. The foundation should be wide enough to increase damping in the rocking mode. The width should be at least (1-1.5) times the vertical distance from the base to machine centreline.

6. The combined center of gravity should coincide with the center of gravity of the foundation.

7. For large reciprocating machines, it may be desirable to increase the embedded depth in soil such that 50% to 80% of the depth, this will increase the lateral restrain and damping ratio for all modes of vibration.

8. Static bearing capacity qall : proportion of footing area for 50% of allowable soil pressure,

which means that the actual soil pressure should be less than 50% of static bearing capacity q_{all}. The actual soil pressure equals to the weight of machine and foundation divided by the base area of footing as shown:

$$\text{Actual soil pressure} = \frac{W_{mach.} + W_{fou.}}{L_f \cdot B_f}$$

9. Static settlement must be uniform; center of gravity of footing and machine load should be within 5% of each linear dimension from the foundation center.

10. Bearing capacity: static plus dynamic loads. The sum of static and modified dynamic loads should not create bearing pressure greater than 75% of allowable soil pressure given for static load condition q_{all}.

11. The magnification factor (M) should preferably be less than (1.5). The magnification factor can be defined as the ratio of dynamic displacement to the static displacement as shown in Table.

12. Vibration amplitude (Y), at operating frequency is shown in figure below that depicts horizontal Amplitude of vibrations. The maximum amplitude of motion for the foundation system should lie in zones A or B.

13. The velocity which equals ($2\pi f_X$ displacement amplitude) compares with the limiting value in table and figure below that depicts horizontal Amplitude of vibrations.

14. The acceleration which equals $4\pi^2 f^2{}_X$ displacement amplitude) should be tested for zone B in figure below that depicts horizontal Amplitude of vibrations.

F=Operating speed of machine $= \dfrac{\omega}{2\pi}$

15. Resonance: the acting frequencies of machine should have at least a difference of ±20 % with the resonance frequency of table

0.8 fmr ≥ f ≥ 1.2 fmr

16. The horizontal translation and the rocking mode needs not be coupled if:

$$\sqrt{f^2{}_{nx} + f_{n\psi}{}^2} / f_{nx}\left(f_{n\psi} \leq 2/3f\right)$$

where:

$f^2{}_{nx=}$natural frequency in the x- direction, rpm.

$f^2{}_{n\psi}$ =natural frequency in the rocking direction, rpm.

Table: Summary of derived expressions for a single-degree-of-freedom system.

Expression	Constant Force Excitation F_o Constant	Rotating Mass-type Excitation $F_o = m_i e w^2$
Magnification factor	$M = \dfrac{1}{\sqrt{(1-r^2)+(2Dr^2)}}$	$M = \dfrac{1}{\sqrt{(1-r^2)+(2Dr^2)}}$
Amplitude frequency f	$Y = M(F_o/k)$	$Y = M_r(m_i e/m)$
Resonance frequency	$f_{mr} = f_n = \sqrt{1-2D^2}$	$f_{mr} = \dfrac{f_n}{\sqrt{1-2D^2}}$
Amplitude at resonance frequency f_r	$Y_{max} = \dfrac{F_o/k}{2d\sqrt{1-D^2}}$	$Y_{max} = \dfrac{(m_i e/m)}{2D\sqrt{1-D^2}}$
Transmissibility factor	$T = \dfrac{\sqrt{1+(2Dr^2)}}{(1-r^2)+(2Dr^2)}$	$\bar{T}_r = \dfrac{r^2\sqrt{1+(2Dr^2)}}{(1-r^2)+(2Dr^2)}$

where:

$r = \omega/\omega^n$

ω_n = Natural circular frequency rad / sec.

ω = requency of excitation force

k = Spring constant, kN /m

m = Mass of machine and foundation, kg

m_i = Rotating mass, kg

D = Damping ratio $= C /C_c$

C = Damping

C_c = Critical damping $= 2\sqrt{km}$

e = Eccentricity of unbalance mass to axis of rotation at operating speed, m

f_n = Natural frequency, rpm

f_{mr} = Resonant frequency for rotating mass-type excitation, rpm

M = Dynamic magnification factor

M_r = Magnification factor

F_o = Amplitude of excitation force, kN

T_r = Force transmitted / F_o

$\bar{T}_r$ = Force transmitted/ $m_i e \omega^2_n$

Y = Amplitude at frequency f

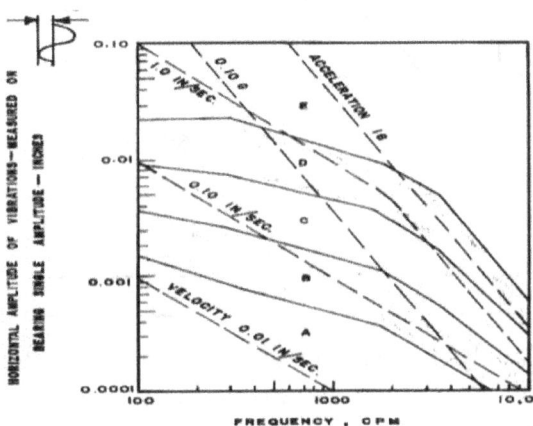

Figure: Vibration performance of rotating machines

    A No faults. Typical new equipment.

    B Minor faults. Correction wasted dollars.

    C Faulty. Correction within 10 days to save maintenance dollars.

    D Failure is near. Correct within two days to avoid breakdown.

    E Dangerous. Shut it down now to avoid danger.

Table: General machinery-vibration-severity data.

Horizontal Peak Velocity (m/sec.)	Machine Operation
< 0.00013	Extremely smooth
0.00013-0.00025	Very smooth
0.00025-0.00051	Smooth
0.00051-0.00101	Very good
0.00101-0.00203	Good
0.00203-0.00406	Fair
0.00406-0.008	Slightly rough
0.008-0.016	Rough
>0.016	Very rough

Design Charts for Machine Foundations

Design charts are prepared to be a guide for the designer engineer. The selected values used in these charts were limited based on the conditions considered in table below as well as the limitations considered in the limitations of machine foundation. The design charts are selected based on three displacements which are acceptable for design of machine foundations as considered in figure below.

Table: The parameters of the empirical method.

Parameters	Basic values	Range of values	Units
$W_{mach.}$	1444.905	60-620	kN
f	585	50-1000	rpm

γ	18.33	18-22	kN/m³
G	96365	25000-190000	kN/m²
ν	0.35	0.3-0.45	-
D_i	0.05	0.05-0.15	---
L_f	8.39	2-20	m
B_f	4.80	2-20	m
h	1.52	0.6-2.2	m
W_{fou}	1443.95	$(3-5)W_{mach}$	kN

For these displacements, the analysis is carried out using the computer program MATHCAD and the results are presented in the form of a relationship between $(G /γ L_f)$ (y-axis) and frequency (rpm) (x-axis), for different ratios of the weight of the foundation to the weight of the machine $((W_f / W_m) W_f$ = weight of foundation, W_m = weight of machine) ranging between (3-5).

The selected displacement values ranged between $(2.5 x 10^{-6})$ m to a maximum value of $(125 x 10^{-6})$ m.

The charts are used to design the dimensions of the footing by the empirical method depending on the weight of the machine, the operating frequency of the machine and the properties of the soil including (shear modulus, Poisson's ratio and unit weight of the soil). In this paper we will take the effects of the minimum displacement = $(2. 5 x 10^{-6}$ m) on the design charts.

This displacement is considered for general limits of vibration which is not noticeable to persons as shown in figure below is drawn for the foundation dimensions ratio L_f /B_f = 1, Poisson's ratio, ν = 0.35, and different soil unit weights, γ = 18, 20 and 22 kN /m³.

From these figures, it is apparent that the frequency is inversely proportional to the values of $(G /γ L_f)$. The curves of these relationships for different values of (W_f / W_m) coincide with each other especially at frequency level (500-1750 rpm). After this limit of frequency, the effect of the weight ratio can be pronounced.

The values of the shear modulus (G) used in these figures ranged between

$(25 x 10^3$ and $175 x 10^3)$ kN/m².

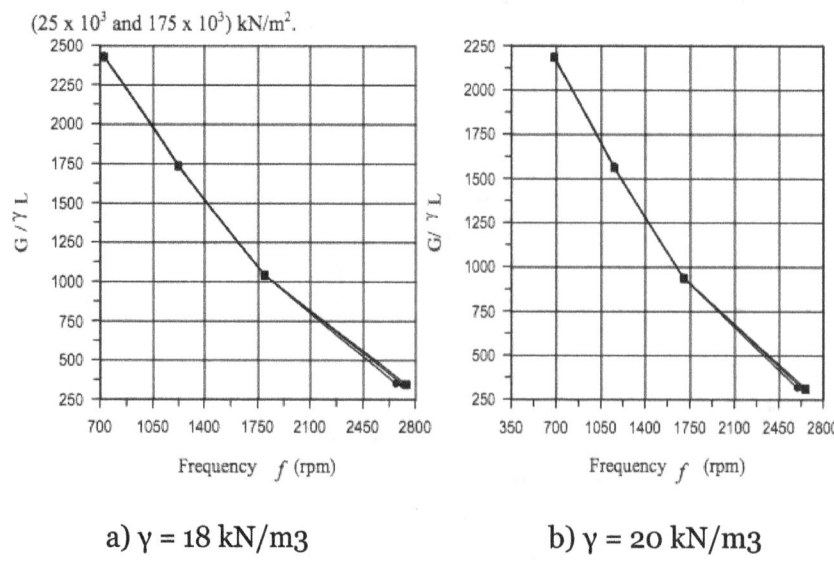

a) γ = 18 kN/m3   b) γ = 20 kN/m3

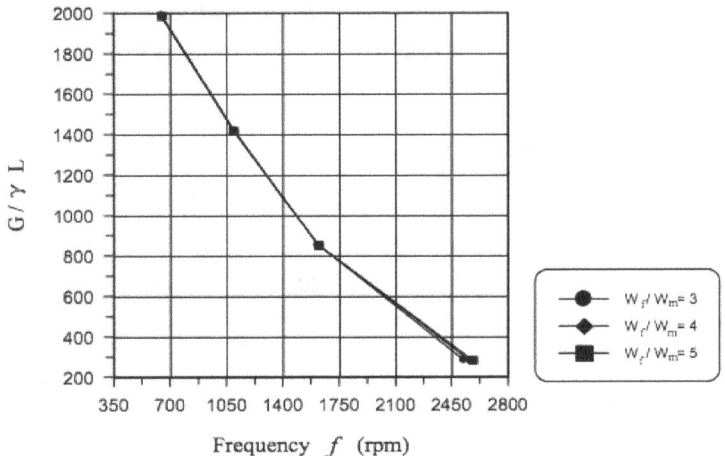

Figure is drawn for the foundation dimensions ratio $L_f/B_f = 2$, Poisson's ratio, $v = 0.35$, and different soil unit weights, $\gamma = 18, 20$ and 22 kN /m³.

a) $\gamma = 18$ kN/ m³

b) $\gamma = 20$ kN/ m³

c) $\gamma = 22$ kN/ m³

From these figures, it is apparent that the frequency is also inversely proportional to the values of $(G /_\gamma L_f)$. The curves of these relationships for different values of (W_f / W_m) coincide with each other especially at frequency level (200-900 rpm). After this limit of frequency, the effect of the weight ratio can be pronounced.

The values of the shear modulus (G) used in these figures ranged between (25 x 10³ and

125 x 10³) kN/m² because in the case of shear modulus equals to (175 x 10³) kN/m², the resulted displacements were out of the limit of (2. 5 x 10⁻⁶) m.

Figure is drawn for the foundation dimensions ratio $L_f/B_f = 3$, Poisson's ratio, $v = 0.35$, and different soil unit weights, $\gamma = 18, 20$ and 22 kN /m³.

From these figures, it is apparent that the frequency is inversely proportional to the values of $(G /_\gamma L_f)$. The curves of these relationships for different values of (W_f / W_m) coincide with each other especially at frequency level (200-1750 rpm). After this limit of frequency, the effect of the weight ratio can be pronounced.

The values of the shear modulus (G) used in these figures ranged between (25 x 10³ and 175 x 10³) kN/m² except for $\gamma = 18$, the shear modulus (G) ranged between (25 x 10³ and 125 x 10³)kN/m² .

Figure is drawn for the foundation dimensions ratio $L_f/B_f = 1$, Poisson's ratio, $v = 0.4$, and different soil unit weights, $\gamma = 18, 20$ and 22 kN /m³.

As in the previous figures, it is apparent that the frequency is inversely proportional to the values of $(G /\gamma L_f)$. The curves of these relationships for different values of (W_f / W_m) coincide with each other especially at frequency level (400-2000 rpm). After this limit of frequency, the effect of the weight ratio can be pronounced.

The values of the shear modulus (G) used in these figures ranged between (25 x 10³ and 125 x 10³) kN/m² except for γ = 18 kN/m³ , the shear modulus (G) ranged between (25 x 10³ and 175 x 10³) kN/ m².

Figure is drawn for the foundation dimensions ratio L_f /B_f = 2, Poisson's ratio, ν = 0.4, and different soil unit weights, γ = 18, 20 and 22 kN /m³.

The same relationship between the frequency and the values of (G /γ L_f). The curves of these relationships for different values of (W_f / W_m) coincide with each other especially at frequency level (400-1000 rpm). After this limit of frequency, the effect of the weight ratio can be pronounced.

The values of the shear modulus (G) used in these figures ranged (25 x 10³and 125 x 10³)

     kN/ m².

Figure is drawn for the foundation dimensions ratio L_f /B_f = 3, Poisson's ratio, ν = 0.4, and different soil unit weights, γ = 18, 20 and 22 kN /m³. From these figures, it is apparent that the frequency is inversely proportional to the values of (G /γ L_f). The curves of these relationships for different values of (W_f/ W_m) coincide with each other especially at frequency level (500-700 rpm). After this limit of frequency, the effect of the weight ratio can be pronounced.

The values of the shear modulus (G) used in these figures are (25 x 10³ and 75 x 10³) kN/m².

Figure below is drawn for the foundation dimensions ratio L_f /B_f = 1, Poisson's ratio, ν = 0.45, and different soil unit weights, γ = 18, 20 and 22 kN /m³.

From these figures, it is apparent that the frequency is inversely proportional to the values of (G /γ L_f). The curves of these relationships for different values of (W_f / W_m) coincide with each other especially at frequency level (250-2000 rpm). After this limit of frequency, the effect of the weight ratio can be pronounced.

The values of the shear modulus (G) used in these figures ranged between (25 x [103] and 175 x 10³) kN/ m².

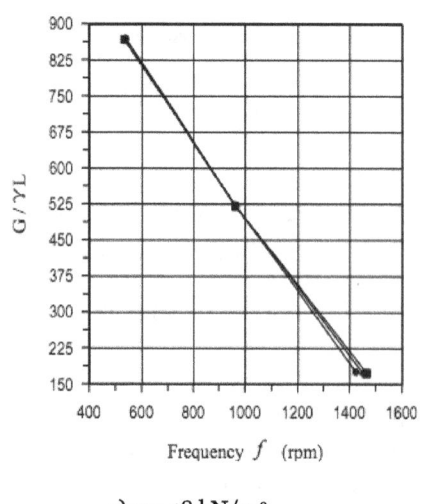

a) γ = 18 kN/m³

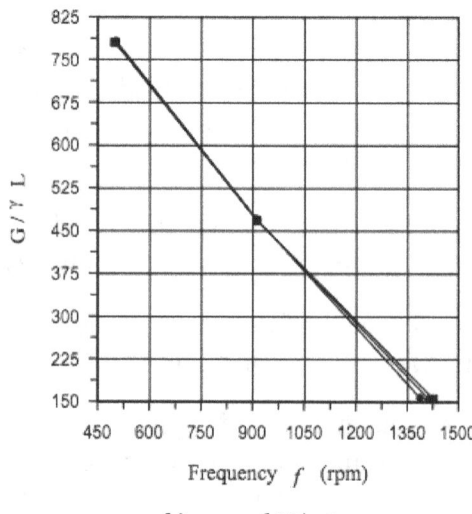

b) γ = 20 kN/m³

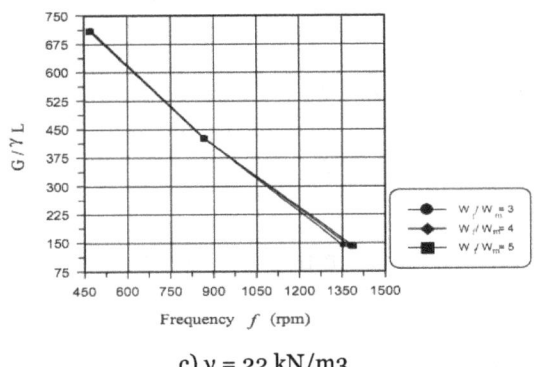

c) γ = 22 kN/m3

Figure: Design charts for machine foundations (L/B = 2, ν = 0.35) and displacement = 2.5 x 10-6 m.

Figure above is drawn for the foundation dimensions ratio $L_f/B_f = 2$, Poisson's ratio, ν = 0.45, and different soil unit weights, γ = 18, 20 and 22 kN /m³.

From these figures, it is apparent that the frequency is inversely proportional to the values of (G /γ L_f). The curves of these relationships for different values of (W_f / W_m) also coincide with each other especially at frequency level (200-1000 rpm). After this limit of frequency, the effect of the weight ratio can be pronounced.

The values of the shear modulus (G) used in these figures ranged between (25 x 10³ and 75 x103) kN/m².

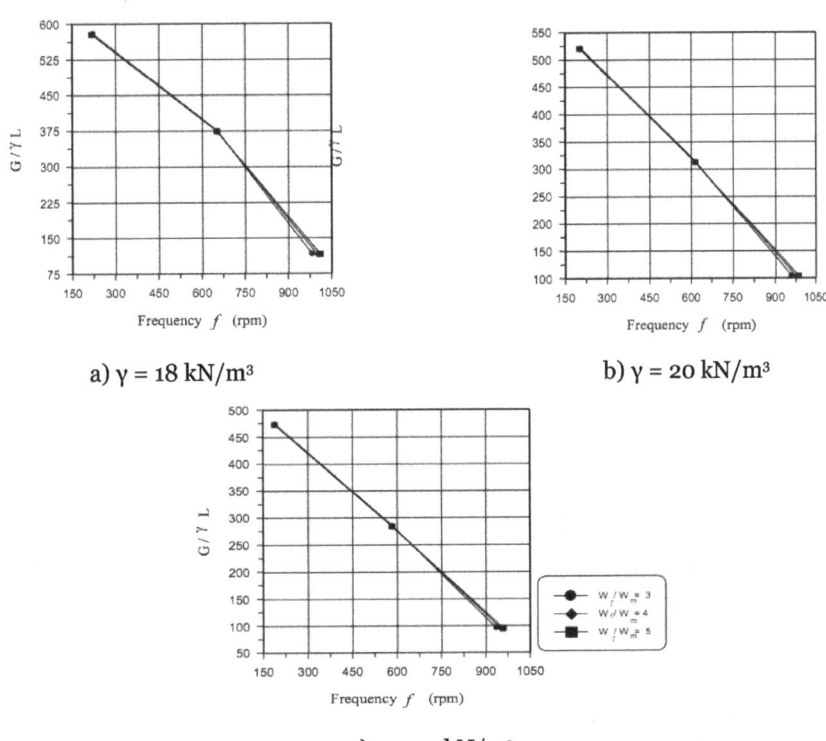

a) γ = 18 kN/m³

b) γ = 20 kN/m³

c) γ = 22 kN/m³

Figure below design charts for machine foundations (L/B = 3, ν = 0.35) and displacement = 2.5 x 10⁻⁶ m.

Figure is drawn for the foundation dimensions ratio $L_f/B_f = 3$, Poisson's ratio, $v = 0.45$, and different soil unit weights, $\gamma = 18$, 20 and 22 kN /m³.

From these figures, it is apparent that the frequency is inversely proportional to the values of (G /γ L$_f$). The curves of these relationships for different values of (W_f / W_m) coincide with each other especially at frequency level (500-650 rpm). After this limit of frequency, the effect of the weight ratio can be pronounced.

The values of the shear modulus (G) used in these figures are (25 x 10³ and 75 x 10³) kN/m²

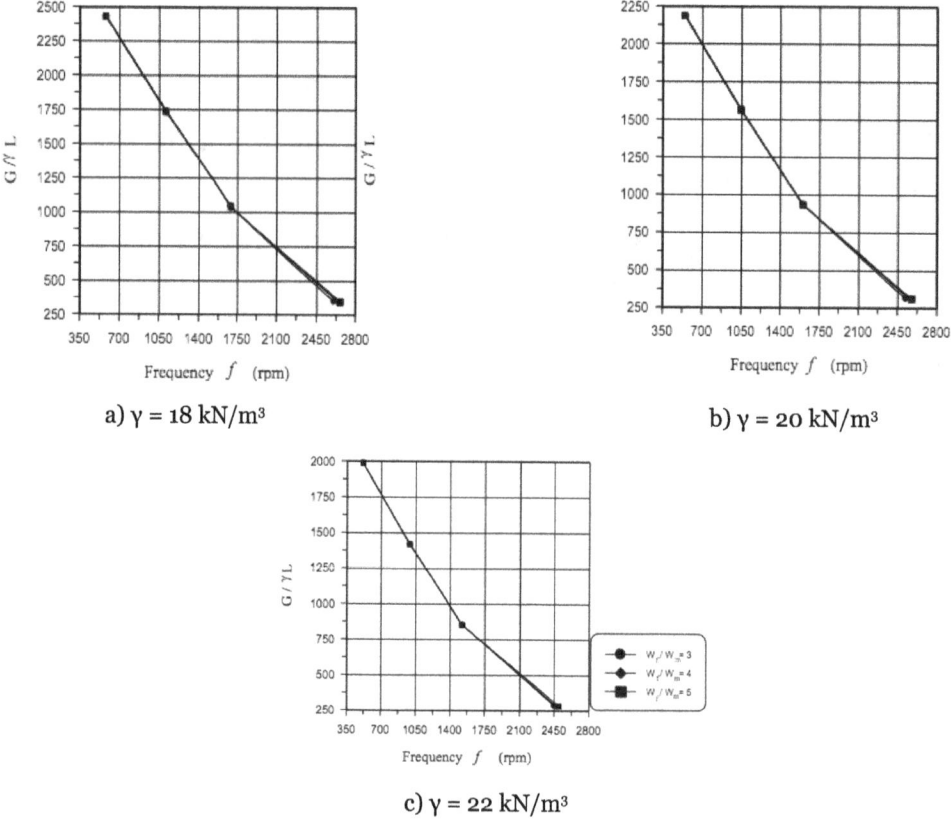

a) γ = 18 kN/m³   b) γ = 20 kN/m³

c) γ = 22 kN/m³

Figure: Design charts for machine foundations (L/B = 3, v = 0.4) and displacement =2.5 x 10⁻⁶ m.

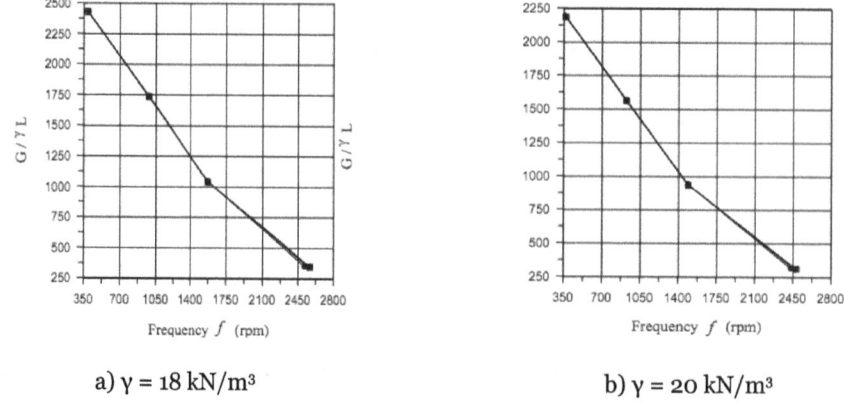

a) γ = 18 kN/m³   b) γ = 20 kN/m³

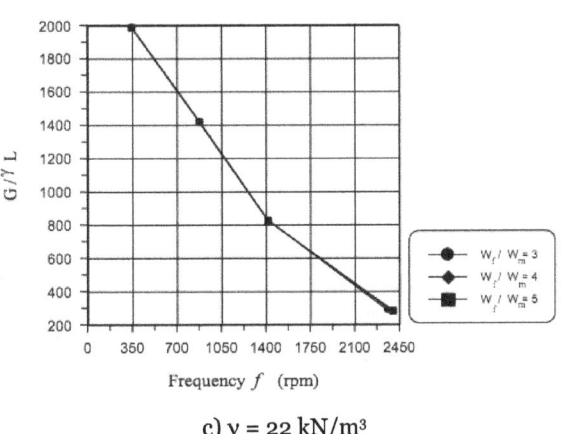

c) γ = 22 kN/m³

Figure: Design charts for machine foundations (L/B = 1, v = 0.45) and displacement =2.5 x 10⁻⁶ m.

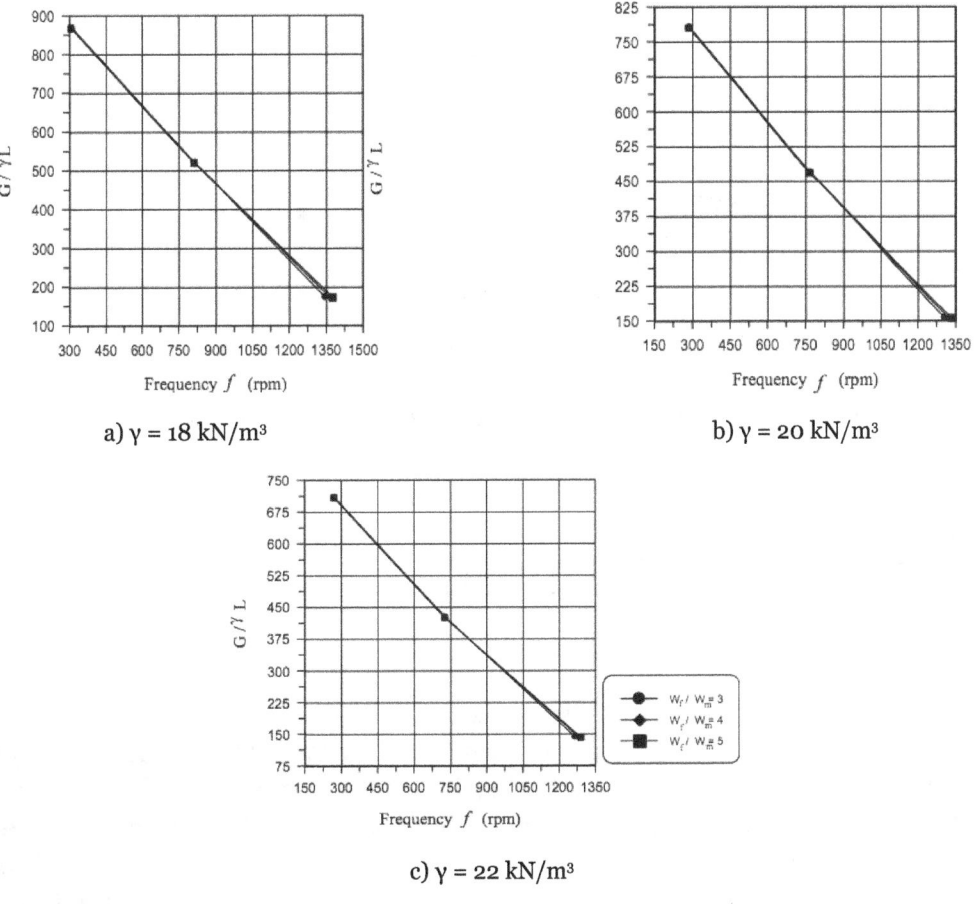

a) γ = 18 kN/m³

b) γ = 20 kN/m³

c) γ = 22 kN/m³

Figure: Design charts for machine foundations (L/B = 2, v = 0.45) and displacement = 2.5 x 10⁻⁶ m.

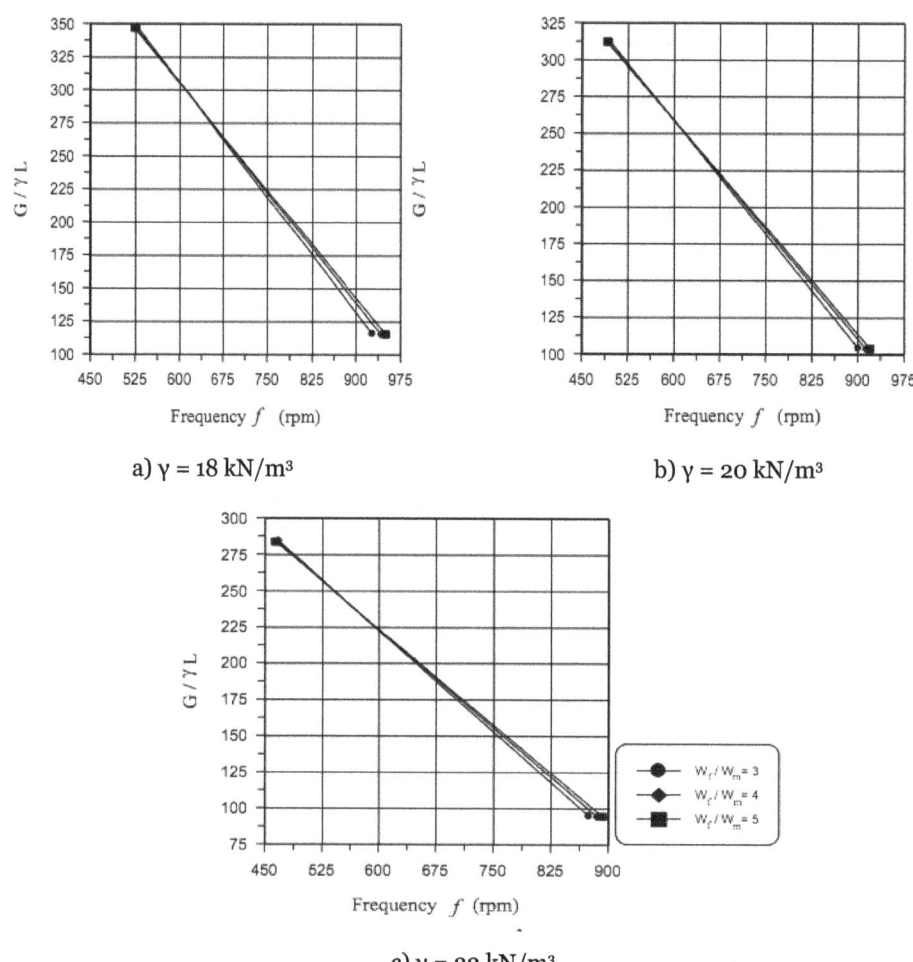

a) γ = 18 kN/m³

b) γ = 20 kN/m³

c) γ = 22 kN/m³

Figure: Design charts for machine foundations (L/B = 3, ν = 0.45) and displacement = 2.5 x 10-6m

Machine Foundation Analysis and Design

General Design Requirements

Foundations shall be cast on a 50 mm thick lean concrete layer. Top of concrete foundations shall be at least 200 mm above the high point of paving. In order to allow the adjustment of pumps exchangers, drums, columns, steel structures etc., during the casting of the foundation, the top of piers shall be left at least 25 mm below the final level. Surfaces under base plates for equipment and structures shall be rough to increase grout bonding. Non shrinking grout shall be used for filling between concrete foundation and baseplates. For large base plates this minimum shall be increased up to 50 mm. Concrete bases for steel plates shall extend at least 25 mm beyond the border of the plate. Minimum distance between the axis of the anchor bolt and the foundation edge shall be:

   120 mm for 14 <= f <= 24 mm

150 mm for 27 <= f <= 45 mm

200 mm for f > 45 mm

Where f = bolt diameter

Anchor bolts shall be positioned within the reinforcing bar cage.

As a general rule anchor bolts shall be installed before concrete casting. If necessary, adequate pockets shall be provided in the foundation where anchor bolts can be installed later. Pockets shall be filled using non-shrinking grout.

Foundation for vessels on skirt shall have the area within the skirt sloped for drainage and an embedded pipe or opening in the grout discharging outside the skirt.

Minimum concrete cover shall be:

Exposition (Minimum concrete cover)

Concrete cast against and permanently exposed to earth.	(70 mm)
Concrete exposed to earth but cast in forms	(50 mm)
Concrete exposed to weather	
For bars <= 16 mm	(40 mm)
For bars > 16 mm	(50 mm)
Concrete no exposed weather nor in contact with ground:	
Walls and slabs	(20 mm)
Beams and Columns	(40 mm)
Shells, folded plate members:	
For bars <= 16 mm	(15 mm)
For bars > 16 mm	(20 mm)

Minimum distance between reinforcing bars shall not be less than 1 time the maximum diameter of bar or 25 mm.

Minimum sliding safety factor shall be 1.50, assuming the coefficient of friction given in the soil report. With the exception of those cases in which the horizontal force depends on the weight of soil (e.g.: retaining walls), such weight shall not be considered if it increases stability.

Minimum overturning safety factor shall be 1.50 during erection and 1.8 in operation. 100% of foundation base area shall be in compression during operation. During erection such percentage may be reduced to 85%.

Foundations and Structures for Vibrating Machinery

- Scope

The following mandatory requirements shall govern the design and testing of supporting structures and foundations for heavy machinery. Heavy machinery is any equipment having reciprocating or rotary masses as the major moving parts (such as reciprocating or rotary compressors, horizontal pumps, engines and turbines) and having a gross plan area of more than 2.5 m² or a total weight greater than 25 KN.

- Design criteria for all heavy machinery

Dynamic modules of elasticity (E') in MPa for use in the dynamic analysis shall be as indicated in DIN 1045. Modulus of elasticity (E) or shear modulus (G) of soil to be used in dynamic shall be as indicated in the soil report. Soil bearing pressure shall not exceed 50% of the net allowable values for static loads. Shrinkage and thermal expansion effects shall be taken into account. In order to prevent Cracking, minimum concrete reinforcing shall be 50 kg/m cubic meter. All reinforcement shall be triaxially arranged.

The Following rules shall be considered in foundation design:

- Foundation design shall consist of clean simple lines;
- Pockets where vapors could accumulate shall not be permitted;
- beams and columns shape should be uniform and rectangular;
- Slender elements shall not be used.

All parts of machine supports shall be independent from the adjacent foundations and buildings. Concrete floor slabs, adjacent to machine foundations, shall be spaced a minimum of 10 mm from the foundations. The space between slab and foundation shall be filled with a flexible joint filler and sealer.

The thickness of the foundation slab, in meters, shall not be less than:

Thk = 0.6 + L / 30

Where:

For one machinery train:

L = longest dimension of the foundation slab (m);

For two or more machinery trains supported by a common foundation:

L = the greater of width of the common slab;

Maximum slab segment length assigned to any train.

In any case minimum thickness of foundation slab shall not be less than 1/10 of its maximum dimension.

The height of supports above grade shall be the minimum required to accommodate suction and discharge piping configuration.

Design Criteria for Reciprocating Machinery

Design foundation for reciprocating machinery shall be carried out in accordance with the following criteria:

- Total foundation weight shall be at least 5 times the total machinery weight.

- Horizontal eccentricity in any direction between the centroid of mass of the machine-foundation system and the centroid of the base contact area shall not exceed 5% of the respective base dimension.

- The center of gravity of the machine-foundation system should be as close as possible to the lines of action of the unbalanced forces.

- Compressor foundations shall include integral supports for the pulsation bottles.

- Groups of reciprocating machinery could be tied together with a common foundation slab when allowed by their location and service.

Dynamic design shall be as follows :

Natural frequencies in the modes being excited shall preferably be out of the range of 0.7 to 1.3 times the disturbing frequencies of any machine on the foundation. If it is not possible fulfill this prescription, frequency within the above mentioned range can be accepted if the maximum amplitudes shall within the limits listed in the following point:

- Damping shall not be higher than 5%.

- Primary forces, couples and moments shall be applied at machine speed to calculate primary amplitudes.

- Secondary forces, couples and moments shall be applied at twice the machine speed to calculate secondary amplitudes.

- Total amplitude shall be calculated by combining, in phase, primary and secondary amplitudes. Total peak-to-peak amplitude on the foundation shall exceed 0.05 mm.

Design Criteria for rotary Machinery

Rotary machinery may be supported either on a direct foundation or an elevated structure. Structure and foundation supporting rotary machines shall be designed in accordance with the requirements of DIN 4024. Foundation and elevated structures shall be dimensioned applying the criteria mentioned in this specification and in accordance with the prescriptions included in par. 7 of DIN 4024. Furthermore weight of basement (foundation and elevated structure) shall be at least 3 times the weight of the machinery.

Elevated structures for rotary machinery shall be as follows:

- Machinery loads shall be directly over vertical supports, where possible.

- Within the weight requirements of the foundation, the upper table and the foundation slab shall be as rigid as possible in the horizontal plane.

- The weight of the foundation slab shall not be less than the combined supported weight of the upper table, columns, walls and machines.

Static design for all types of foundations shall take into account the following loads:

- Dead weight of machines and their base plates.

- Transversal forces representing 25% of the weight of each machine, including its baseplate, applied normal to its shaft at a point midway between the end bearings.

- Longitudinal forces representing 25% of the weight of each machine, including its baseplate, applied along the shaft axis.

- Total transversal and total longitudinal forces per b. and c. above shall not be considered to act concurrently.

Dynamic design for all types of foundations shall be as follows:

- Barkanś theory shall be utilized to carry out the calculations of natural frequencies and amplitudes.

- All natural frequencies shall be out of the range of 0.7 to 1.4 times operating speed of any machine supported thereon; where this frequency ratio restriction is impractical or uneconomical, frequency ratios within the above range will be accepted, provided the amplitudes meet the requirements of point.

- Short circuit couples, oil whirl frequency, rotor critical speeds and background vibration shall also be considered.

- Transverse bents or walls should be designed so that their vertical natural frequencies agree within 5%.

- Torsional, transverse and longitudinal horizontal natural frequencies should be determined considering the whole structure. Individual transverse bents or walls should have same transverse horizontal frequencies.

- Multi-degrees of freedom shall be considered if a single degree of freedom system will not lead to a reasonable mathematical representation of the structure.

- Loaded beam, slab and frame natural frequencies in both horizontal and vertical directions, where possible, shall be above any machine speed. If beams, slabs or frames must be designed to have natural frequencies below machine speed, allowance must be made for the stiffening effect of the base plate and the machine.

Amplitudes shall be determined using dynamic forces from each rotor, calculated as follows:

Dynamic force = Rotor weight x Rotor Speed (rpm).

- Total amplitude on the structure of foundation in any direction shall not exceed the values indicated in the following table:

ALLOWABLE AMPLITUDES

ROTOR SPEED (RPM)	PEAK TO PEAK AMPLITUDE (MM)
0-999	0.023
1000-1149	0.020
1150-1299	0.018
1300-1499	0.015
1500 and above	0.013

Design Criteria of Light Vibrating Machinery

Following mandatory requirements shall govern the design of supporting structures and foundations for light vibrating machinery:

- Light vibrating machinery is any equipment having reciprocating or rotary masses as the major moving parts (such as reciprocating or rotary compressors, horizontal pumps etc.) and having both a gross plan area less than 2.5 m² and a total weight less than 25 KN.

- For light vibrating machinery dynamic design shall be neglected. Static design of foundations shall be performed according to clause 7.4 of this specification, but weight of foundation must be at least 3 times the total rotary machines weight or 5 times the total reciprocating machines weight.

- Minimum concrete class shall be f´c= 21 N/mm².

Methods of Analysis

The analysis of machine foundation is usually performed by idealizing it as a simple system. Figure shows a schematic sketch of a rigid concrete block resting on the ground surface and supporting a machine. Let us assume that the operation of the machine produces a vertical unbalanced force which passes through the combined centre of gravity of the machine-foundation system. Under this condition, the foundation will vibrate only in the vertical direction about its mean position of static equilibrium. The vibration of the foundation results in transmission of waves through the soil. These waves carry energy with them. This loss of energy is termed geometrical damping". The soil below the footing experiences cyclic deformations and absorbs some energy which is termed, material damping". The material damping is generally small compared to the geometrical damping and may be neglected in most cases.

However, material damping may also become important in some cases of machine foundation vibrations. The problem of a rigid block foundation resting on the ground surface, may therefore be represented in a reasonable manner by a spring-mass-dashpot system shown in Fig.below. The spring in this figure is the equivalent soil spring which represents the elastic resistance of the soil below the base of the foundation. The dashpot represents the energy loss or the damping effect. The mass in figure b is the mass of the foundation block and the machine. If damping is neglected, a spring-mass system shown in figure c may be used to represent the problem defined in figure. a.

Single degree of freedom models shown in figure b and c may in fact be used to represent the problem of machine foundation vibration in any mode of vibration if appropriate values of equivalent soil spring and damping constants are used. For coupled modes of vibration, as for combined rocking and sliding, two degree-of-freedom model is used.

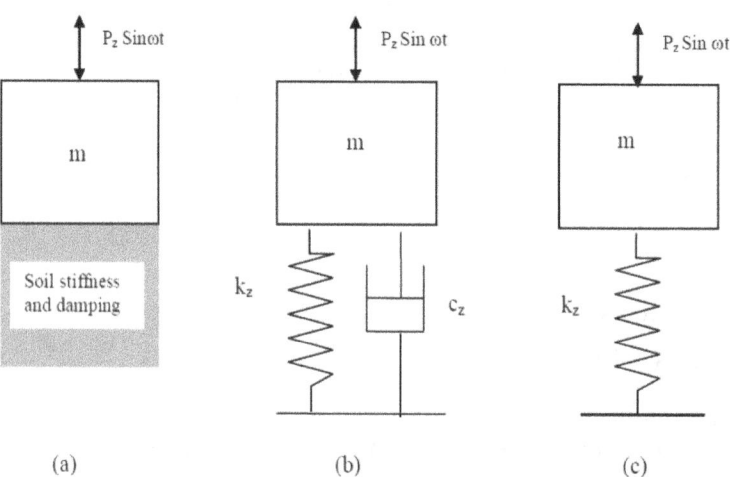

Vertical Vibrations of a Machine Foundation (a) Actual Case (b) Equivalent model with damping
(c) Model without damping

All foundations in practice are placed at a certain depth below the ground surface. As a result of this embedment, the soil resistance to vibration develops not only below the base of the foundation but also along the embedded portion of the sides of the foundation. Similarly the energy loss due to radiation damping will occur not only below the foundation base but also along the sides of the foundation. The type of models shown in figures b and c may be used to calculate the response of embedded foundations if the equivalent soil spring and damping values are suitably modified by taking into account the behavior of the soil below the base and on the sides of the foundation.

Similarly the energy loss due to radiation damping will occur not only below the foundation base but also along the sides of the foundation. The type of models shown in figures b and c may be used to calculate the response of embedded foundations if the equivalent soil spring and damping values are suitably modified by taking into account the behavior of the soil below the base and on the sides of the foundation.

For designing new foundations or retrofitting existing foundations for machines, calculate the natural frequency of the machine-foundation-soil system so as to avoid resonance, evaluate the amplitude of vibrations of the machine-foundation-soil system under operating dynamic loads and frequencies and make sure that they satisfy the acceptability criteria. with reference to avoiding damage to machinery – it means amplitude of velocity with the limiting value prescribed by the manufacturer and in most cases, it will be governing criterion and will dictate your design. with reference to avoiding damage to building structures- it can be either amplitude of displacement or velocity as prescribed by local codes for different types of construction, with reference to ensuring comfort of persons-it means amplitude of displacement usually prescribed by the local codes for different frequencies of operation, and with reference to avoiding excessive settlement due to large number of cycles of load applications- it means amplitude of displacement prescribed by the local codes particularly for soil profiles consisting of loose sands and silts with high water table.

In order to calculate the natural frequency and amplitude of vibrations for a particular machine-foundation-soil-system, you need to know the local soil profile and soil characteristics as also the dynamic loads generated by the machine that are provided by the manufacturer.

1. Empirical

2. Elastic half space method

3. Linear elastic weightless spring method.

Empirical Method

On such design guideline, rather a rule of thumb was the weight of the foundation should be at least three to five times the weight of machine being supported. There are some empirical formulae available in literature for estimating the natural frequency, mostly for the vertical mode of vibration. In these formulae, it is assumed that a certain part of the soil, immediately below the foundation, moves as a rigid body along with the foundation and is called apparent soil mass or in-phase mass. For example, D.D.Barken in 1962 suggested that the mass of the vibrating soil should be between 2/3 to 3/2 times the weight of foundation and machine. These guidelines/formulae do not take into account the nature of subsoil, type of excitation force (harmonic/impact), contact area and mode of vibration.

In using the second and third methods, we use the concept of discrete system or lumped parameter system, In this system: the mass of the machine and the supporting foundation are lumped together as one discrete rigid mass 'm'- this is reasonable since the stiffness of reinforced concrete and the machine is several times more than that of soil. the soil resistance is represented by means of a spring with a constant 'k' ,and the energy absorption characteristic of soil is represented by means of a dashpot with a constant 'c'. Once we have these three quantities m, k, and c, using the theory of vibrations, the natural frequency as well as the amplitude of vibration for any given type of load can be calculated. The formulae to calculate amplitude of vibration for either impact or cyclic type of loading can be found in books dealing with structural dynamics.

The soil behaves essentially as a linear elastic material and the resistance to deformation offered by it can be represented by a spring. This is based on the fact that the magnitude of the dynamic force acting on the foundation is very small compared to the static load, normally no more than 10%. And, therefore, the amplitude of displacement is extremely small. Since the non-linear behavior of the soil does not come into play, material dumping in the soil can be neglected. In machine foundations, however, there exists another kind of damping known as radiation damping or geometric damping. Every time a dynamically loaded foundation moves against soil, stress waves originate at the contact surface in the form of surface waves and carry away some of the energy transmitted to the soil.

This loss of energy is called radiation damping and in passing that it can be effectively modeled by a dashpot. Use of the elastic half space method or the linear elastic weightless spring method enables us to calculate the value k and c. number of ways or modes in which a machine foundation can deform. This block foundation can deform in any one or all of six possible modes.

Degrees of Freedom of a Rigid Block Foundation

A typical concrete block is regarded as rigid as compared to the soil over which it rests. Therefore, it may be assumed that it undergoes only rigid-body displacements and rotations.

Any rigid-body displacement of the block can be resolved into these six independent displacements. Hence, the rigid block has six degrees of freedom and six natural frequencies. Of six types of motion, translation along the Z axis and rotation about the Z axis can occur independently of any other motion. However, translation about the X axis (or Y axis) and rotation about the Y axis (or X axis) are coupled motions.

Therefore, in the analysis of a block, we have to concern ourselves with four types of motions. Two motions are independent and two are coupled. For determination of the natural frequencies, in coupled modes, the natural frequencies of the system in pure translation and pure rocking need to be determined. Also, the states of stress below the block in all four modes of vibrations are quite different. Therefore, the corresponding soil-spring constants need to be defined before any analysis of the foundations can be undertaken.

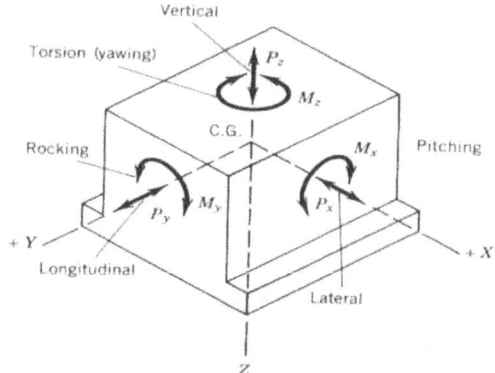

Modes of Vibration of a rigid block foundation

Elastic Half Space Method

This method is called the elastic half space method because the ground is assumed to be an elastic, homogeneous, isotropic, semi-infinite body which in the Theory of Elasticity is referred to as elastic half space. The elastic half space theory can be used to determine the values of equivalent soil springs and damping then make use of theory of vibrations to determine the response of the foundation. These are known as the "the elastic half space analogs". The machine foundation is idealized as a mechanical oscillator with a circular base resting on the surface of ground.The Boussinesq's solution for finding the stresses induced in soil due to a point load on the ground surface. In the elastic half space method, the point load is assumed to be dynamic. By integrating the solution for a dynamic point load over a circular area, the stresses due to a circular machine foundation is calculate. It may be mentioned here that the equivalent soil spring and damping values depend upon the:

(i) Type of soil and its properties,

(ii) Geometry and layout of the foundation,

(iii) Nature of the foundation vibrations occasioned by unbalanced dynamic loads.

Using the theory the values k and c are calculated. Soil properties that are required to determine k and c are the shear modulus G, mass density ρ and Poisson's μ.

According to the theory with the vertical vibrations of a machine foundation of radius r_0

$$K_z = \frac{4Gr_0}{1-\mu}$$

$$C_z = \frac{3.4r_0^2}{1-\mu}\sqrt{p^G}$$

Linear Elastic Weightless Spring Method

The linear elastic weightless spring method differs from the elastic half space method in two respects: the soil below the foundation is considered weightless – this assumption is not valid but it does not appreciably affect the computed response, and damping in the soil below the foundation is neglected- this assumption is also not valid since radiation damping exists in a machine foundation. The effect of damping is incorporated in this method by independently estimating it using a field test. These assumptions make it possible to represent the machine-foundation-soil-system with and equivalent mass spring system. There is no dashpot here since damping has been neglected. In this method, the spring constant k for vertical vibrations is expressed as a function of the area of the foundation A_f and the coefficient of elastic uniform compression C_u is.

$$K_z = C_u A_f$$

References

- Machine-foundation-its-introduction-and-importance-2264: strukts.com, Retrieved 28 May 2018

- Machine-foundation: dailycivil.com, Retrieved 15 April 2018

- Machine-foundation-vibration-analysis-14777: theconstructor.org, Retrieved 24 June 2018

- General-requirements-of-machine-foundations-7244: theconstructor.org, Retrieved 14 July 2018

- Machine-foundation-analysis-and-design, dynamics, repository: excelcalcs.com, Retrieved 14 April 2018

Permissions

All chapters in this book are published with permission under the Creative Commons Attribution Share Alike License or equivalent. Every chapter published in this book has been scrutinized by our experts. Their significance has been extensively debated. The topics covered herein carry significant information for a comprehensive understanding. They may even be implemented as practical applications or may be referred to as a beginning point for further studies.

We would like to thank the editorial team for lending their expertise to make the book truly unique. They have played a crucial role in the development of this book. Without their invaluable contributions this book wouldn't have been possible. They have made vital efforts to compile up to date information on the varied aspects of this subject to make this book a valuable addition to the collection of many professionals and students.

This book was conceptualized with the vision of imparting up-to-date and integrated information in this field. To ensure the same, a matchless editorial board was set up. Every individual on the board went through rigorous rounds of assessment to prove their worth. After which they invested a large part of their time researching and compiling the most relevant data for our readers.

The editorial board has been involved in producing this book since its inception. They have spent rigorous hours researching and exploring the diverse topics which have resulted in the successful publishing of this book. They have passed on their knowledge of decades through this book. To expedite this challenging task, the publisher supported the team at every step. A small team of assistant editors was also appointed to further simplify the editing procedure and attain best results for the readers.

Apart from the editorial board, the designing team has also invested a significant amount of their time in understanding the subject and creating the most relevant covers. They scrutinized every image to scout for the most suitable representation of the subject and create an appropriate cover for the book.

The publishing team has been an ardent support to the editorial, designing and production team. Their endless efforts to recruit the best for this project, has resulted in the accomplishment of this book. They are a veteran in the field of academics and their pool of knowledge is as vast as their experience in printing. Their expertise and guidance has proved useful at every step. Their uncompromising quality standards have made this book an exceptional effort. Their encouragement from time to time has been an inspiration for everyone.

The publisher and the editorial board hope that this book will prove to be a valuable piece of knowledge for students, practitioners and scholars across the globe.

Index